OLYMPIC NATIONAL PARK

OLYMPIC NATIONAL PARK
A Natural History

TIM McNULTY

UNIVERSITY OF WASHINGTON PRESS
Seattle and London

To Polly Dyer: friend, mentor,
and long-time defender of Olympic wildlands

First published by Houghton Mifflin Company in 1996
Re-issued by Sasquatch Books in 1999
First University of Washington Press edition published, with corrections, in 2003
Printed in the United States of America

Library of Congress Cataloging-in-Publication Data

McNulty, Tim.
    Olympic National Park : a natural history / Tim McNulty—University of Washington Press ed.
        p. cm.
    Includes bibliographical references (p.) and index.
ISBN 0-295-98300-0 (alk. paper)
    1. Natural history—Washington (State)—Olympic National Park. 2. Olympic National Park (Wash.) I. Title.
QH105.W2 M37   2003
508.797'98—dc21

                                                                                    2002035975

The paper used in this publication is acid-free and recycled from 10 percent post-consumer and at least 50 percent pre-consumer waste. It meets the minimum requirements of American National Standard for Information Sciences—Permanence of Paper for Printed Library Materials, ANSI Z39.48-1984.

Book design by Wendy Baylor and Jeremy Schmidt

Acknowledgments

Any attempt to tell the story of a place as wonderfully diverse as Olympic National Park is dependent on both the expertise and generosity of scientists and researchers working in the field. Many knowledgeable individuals took time from their own work to help me understand aspects of the Olympics' natural history, point me toward new research, or review sections of this manuscript. I owe each of them a profound debt of gratitude. Any inaccuracies of course are entirely my own.

At Olympic National Park, research biologist Ed Schreiner generously shared his knowledge of Olympic's plant communities; Bruce Moorhead and Doug Houston shared insights into the park's wildlife; John Meyer clarified issues regarding fisheries; and Erran Seaman shared his research on spotted owls. Paul Gleeson, David Conca, and Jacilee Wray of the cultural resources division shared perspectives on peninsula archaeology and early ethnography. Roger Hoffman assisted with map preparation and contributed insights into bird populations on the peninsula; and Paul Crawford provided valuable background on a number of resource issues.

Special thanks are due to Michael Smithson, chief of interpretation for the park. Michael was extremely supportive of this project from its inception, helped facilitate research at the park, and was always willing to share ideas. Mike Gurling connected me with research on Olympic's coastal ecology; Janis Burger, Sharon Wray, and Susan Schultz assisted with historical photographs, and Janis also shared her knowledge of native plants. Hank Warren offered helpful perspectives on the Olympics' geology.

I am deeply indebted to Jo Davies and Carolyn Kirsch of the Peninsula College Library for their assistance in locating scientific papers and journals. I also wish to thank Ron Crawford and Ed Tisch of Peninsula College; Howard Conway, Glenn Thackray, Joseph Vance, and Andrea Woodward of the University of Washington; Megan Dethier of the University of Washington's Friday Harbor Laboratories; David Shaw of the University of Washington's canopy crane research project; Nalini Nadkarni of the Evergreen State College; Roland Tabor of the U.S. Geological Survey, Jan Henderson and Robin Lesher of the Forest Service Ecology Program; Ulrich Wilson of the U.S. Fish and Wildlife Service; Steve Jeffries, Greg Shirato, Anita McMillan, and Jim Watson of the Washington Department of Fish and Wildlife; Ed Bowlby and Bob Steelquist of the Olympic Coast National Marine Sanctuary; Tom Loughlin, Pat Gearin, and Bob DeLong of the National Marine Fisheries Service Northwest Marine Mammal Laboratory; Barbara Lawrence and Jamie Valadez of the Elwha S'Klallam Tribe; and Mike Reed of the Jamestown S'Klallam Tribe.

I also wish to thank Barbara Blackie, Nelsa Buckingham, Al Charles Jr., Welden Clark, Delbert Gilbrow, Dick Goin, Jerry Gorsline, Ruth Kirk, Gene Kridler, Keith Lazelle, Carsten Lien, Pete Schroeder, Autumn Scott, Fred Sharpe, Patty Swingle, Jimmy Swingle, and Robert Wood for their assistance and support.

Lastly, I would like to thank my wife, Mary Morgan, for her patience, good humor, encouragement, and advice throughout the writing of this book.

CONTENTS

An Island of Rivers

I watched them form as I followed an elk trail high above a timbered basin: light feathery clouds that clung to the peaks as though the mountains had grown wings. Now, hours later, they had thickened and melded into a high plane that spread over the snow-streaked summits of the Olympic Mountains like a quilt. I'd seen the pattern before—the slow dance of mountain and cloud as moist air from the Pacific draws up over the heights of the range—and I knew what was to come.

Letting go of my destination, I climbed to an aerielike camp on a favorite ridge. From there I'd have a splendid view as the pageant of weather once more laid claim to this coastal mountain wilderness. Though I might grouse a bit, I never seriously complain about the weather. It was ocean-borne rain and snow that sculpted these mountains over time, and the ever-present breath of moist marine air continues to stir them to life.

The Olympic Mountains form the backbone of Washington's Olympic Peninsula, that angular thumb of land that juts into the Pacific at the northwest corner of the United States. The heart of the peninsula, more than 1,400 square miles of rugged mountains, richly forested river valleys, and pristine wilderness coast have been preserved as Olympic National Park.

Janis Burger

Mount Olympus becomes shrouded in clouds as moist marine air from the Pacific rolls up the Hoh River valley, rises, and cools. At just under 8,000 feet in height, Mount Olympus receives more than 200 inches of precipitation annually, mostly in the form of snow.

Olympic is one of the premier wilderness parks in the United States, and it harbors one of the richest old-growth forest preserves in the world. The park also protects the core of a larger ecosystem that surrounds it. Simply stated, an ecosystem is a complex of interacting plants and animals large enough to support the habitat needs of all the species it contains. The Olympic ecosystem is a living fabric that stitches the movement of tides over storm-worn coastlines to windblown drifts of snow at the heights of mountain peaks. Within this tapestry lies a wealth of interdependent plant and animal communities flourishing in a landscape of timeless beauty.

Olympic National Park was established a mere half-century ago, but its story reaches back much further than that. The character of the Olympic Mountains, their unique plant and animal species, and the startling natural diversity found here all trace their evolution to the shifts of climate and profound changes in the landscape that culminated during the Pleistocene ice age.

For thousands of years during the Pleistocene, the Olympic Peninsula was isolated from the rest of North America's mountain ranges by continental glaciers that advanced repeatedly from the North. The alpine plant and animal communities that adapted or evolved during this time remained separated from mainland ranges by the inland seas the glaciers left in their wake. Today Puget Sound, the Strait of Juan de Fuca, and the deep valley of the Chehalis River to the south continue to isolate this coastal range, lending an islandlike character to the peninsula's ecosystem. Though not an island in the geographic sense, the Olympic Mountains and the life communities they support can be seen as an island of rivers, separate in a way, but bound to seas and skies, flyways and migratory routes. Lifted up quite recently and weathering down in time, the Olympics are a window into the long and lovely story of the passage of life over the earth, an island where those primal forces still hold sway.

Looking south from my perch, I see the glacier-clad slopes of Mount Olympus rise out of the forested headwaters of the Hoh River in broken sweeps of rock and ice, the summit just tasting the clouds. Ascending nearly 8,000 feet in elevation less than 40 miles from the Pacific, Mount Olympus is the tallest peak in the Olympics. Spilling down its flanks, blue-white in the overcast light, are massive rivers of ice that hearken back to that earlier time of the Pleistocene. These and more than 60 other named glaciers in Olympic serve as a reminder that the 12,000 to 14,000 years that separate us from the last great ice age are a geological blink of an eye. A drop of a few degrees in the earth's mean temperature, and the glaciers could once more reach for the lowland valleys they left such a short time ago.

The glaciers of Mount Olympus visible from my ridge top, the Hoh, the Blue, the White and Ice River glaciers, are fed by more than 200 inches of precipitation each year, mostly in the form of snow. By contrast, less than 34 miles to the northeast, in the rainshadow of the range, Sequim Prairie in the Dungeness valley receives less than 20 inches. The spruce and hemlock forests of the coast give way to pines and junipers in the upper Dungeness, and valley farmers irrigate their crops. The diversity of habitats and plant communities born of this climatic variation contributes to the biological richness of the Olympics and deepens their beauty.

Below me, a tongue of coastal fog has worked its way up the Hoh Valley. Like the ghost of a glacier it rises and subsides against the upper

Morning fog lingers in the Puget Sound lowlands as seen from Deer Park in the northeast Olympics. The Cascade Range in the distance forms the eastern edge of the Puget Sound Basin; Washington's San Juan Islands, in the middle ground, lie within the rainshadow of the Olympic Mountains.

Janis Burger

reaches of forest. As winds stir the fog layer, the muted thunder of the river rises and falls with the cadence of a distant sea. Low clouds and fog, so characteristic of these coastal valleys, are an essential part of the famed temperate rain forests of the Olympics. Beneath the cloud layer, through miles of unbroken valley forest, the moss- and lichen-draped limbs of rain forest trees sift the fog, making a fine rain fall on shaded forest floors. Somewhere in the valley, herds of Roosevelt elk browse the lush understory of ferns and shrubs, and Douglas squirrels cut cones for their winter stores.

For more than a century the rain forest valleys of the Olympics were prized by lumbermen and naturalists alike for their record-size trees.

Janis Burger

Olympic's cool, wet coastal climate nurtures lush temperate rain forests in west-side valleys. By a measure of biomass per acre, these forests are the most productive in the world.

Sitka spruce, western hemlock, western redcedar and Douglas-fir all reach giant proportions here. Only during the past 20 years have scientists begun to understand and appreciate the ecological complexity and richness of these forests and the number of species that depend on them for habitat.

Today, more than 85 percent of the ancient forests that once mantled the lowlands of the Pacific Northwest, including much of the lowland Olympic Peninsula, have been destroyed by logging. But as land managers turn their attention more toward ecological restoration and sustainable resource use, the ancient forests protected by Olympic National Park and the genetic information and ecosystem functions pre-

Janis Burger

More than 60 miles of rugged, undeveloped coastline are protected within the park. Offshore sea stacks and islands provide nesting habitats for thousands of seabirds as well as haul-outs for migratory and resident sea mammals.

served here may prove invaluable. In the meantime, the majestic forests of Olympic continue to offer solace, inspiration, and time-worn perspective to millions of visitors each year.

The isolation of the Olympic Peninsula was a fortunate accident of geography that allowed its forests to survive the 19th-century assault of mining, trapping, logging, and railroad building that altered so much of the western landscape. As a result, the Olympic Peninsula entered the 20th century with its forests, wildlife communities, and watersheds wonderfully intact. To this day no through-roads cross the Olympic Mountains, and some 62 miles of wild, rugged coastline remain free of roads and developments. The rivers that slice their way out from the interior mountains follow their restless courses to the Pacific, the Strait, or Puget Sound largely unfettered. And some valley farmers still have to repair fences after the elk move through.

But all places, even those we treasure most, are but points on a continuum, brief moments in the passage of time through a changing landscape. To appreciate a place like Olympic National Park is to understand the natural processes that created it and brought it to life. It is to understand, too, that the landscape and its life communities will continue to evolve and change. Ancient forests will succumb to windstorms and fire, plant communities to changes in rainfall and climate. Because of their mobility, wildlife populations are most susceptible to activities occurring outside the national park; Olympic's wild salmon runs are affected by logging as well as commercial fishing pressures on the high seas.

Until this century, the land was its own best protector. Early park and national forest designations helped conserve much of the peninsula's resources. But road building and clearcutting have increased dramatically on public as well as private lands in recent years, and pressures to develop rivers, shorelines, and habitat areas persist. As conflicting pressures mount on land managers and public officials, it falls to each of us to assume stewardship for places like Olympic National Park. While natural forces continue to shape the evolution of the Olympic ecosystem, concerned and informed citizens will have increasing say about human-caused impacts on national park resources. Citizen involvement has already transformed the forest service's management of its remaining ancient forests. Informed public opinion has a similar role to play in land use decisions affecting the larger ecosystems that surround and support our national parks.

The native cultures who occupied the Olympic Peninsula for thousands of years lived in dynamic balance with the natural bounty that surrounded them. The coastal cultures of the Pacific Northwest are considered to be among the most sophisticated and highly developed native cultures north of Mexico. Many ancestral native villages still flourish along the Olympic shoreline. They remind us that humans living sustainably with the natural environment have always been a part of the Olympic ecosystem. Their cultures can serve as models as we look for sustainable ways to live and work here today.

As I sip my tea, Mount Olympus and its retinue of peaks have fallen to the descending cloud layer, and valley clouds from the Hoh River rise to join other clouds creeping across ridges to the north and east. Soon I'll be seeing only my boot tips perched on a rock ledge over the void. Rain will be rattling against my tent before dark—already the wind has drowned out the river's distant song—and tomorrow will be much the same. Though I know it's probably futile, I'll give it a day or two before packing up and hiking out. I always do.

As I rummage through my pack for a sweater, two chestnut-backed chickadees land in a shrubby tree beside me, utter their chittery three-note song, puff up and appear to hunker down. It's an old adaptive trait in these mountains: settling in to a turn in the weather. It helps keep us all honest. And when the skies clear, as they eventually do, we all emerge, as from a chrysalis of cloud, and dry our wings in the sparkling light of a world newly made.

1
Shuffled Texts of Stone

The Olympic Mountains must have come as a shock to the early European explorers who sailed the west coast of North America. For a thousand miles the coast ranges rose in gentle folded hills, sere and treeless to the south, progressively greener and more deeply forested as seafarers sailed north. Suddenly, just past the 47th parallel, the coast hills erupted into a high rocky jumble of snowclad summits, ice fields gleaming in the windy light.

Though explorers had plied Pacific Northwest waters since the mid-16th century, the first sighting of the Olympic Mountains wasn't recorded until 1774. On a clear day in August that year, Spanish seafarer Juan Perez sighted the ice-bound summit of the Olympics' highest peak and recorded it in his ship's log as El Cerro de la Santa Rosalia. Fourteen years later, in 1788, British sea captain John Meares spotted the same peak from the Strait of Juan de Fuca to the north. Remote and hidden behind a retinue of ridges and peaks, the mountain struck the captain as a suitable dwelling place for the gods of this new land, and he named it Olympus after the mountain that was home to the gods of Greek mythology. British captain George Vancouver, who explored the inland waters of Puget Sound four years later, extended the name to the range of mountains surrounding Meares's Olympus, and the name held.

Since "Mount Saint Rosalie" was the first geographic feature to be named by Europeans in what is now Washington state, one might suspect the Olympic Mountains would be among the first in Washington to be explored. Instead, the mystery that shrouded the mythic Greek Mount Olympus also obscured this far corner of the New World for most of the next century. It wasn't until the late 1880s, at the close of the American frontier, that systematic explorations of the Olympic Mountains were mounted. Although the native people who lived on the peninsula had developed a rich body of myths recounting the creation of this land, the veil of mystery surrounding the geological origins of the Olympic Mountains would remain until quite recently.

In his 1888 report to the Secretary of the Interior, Washington Territorial Governor Eugene Semple included a section on the Olympic Peninsula:

> The mountains seem to rise from the edge of the water, on both sides, in steep ascent to the line of perpetual snow, as though nature had designed to shut up this spot for her safe retreat forever. . . . In tradition alone has man penetrated its fastness and trod the isles of its continuous woods.

The *Seattle Press* picked up the call for an expedition that would at last unlock "the mystery which wraps the land encircled by the snow capped Olympic range," and the following year, a group backed by the Press began a celebrated exploration of the Olympics. It was not the first party to venture into the heart of the interior Olympics, nor the best organized or equipped, but the members of the Press party achieved lasting notoriety for their dogged and persistent traverse through the heart of the range during one of the worst winters on record. Three grueling months after starting up the Elwha River in December of 1889, expedition leader James Christie and Captain Charles Barnes slogged through deep snow to a ridge crest north of Mount Wilder and at last laid eyes on the interior mountains. Barnes described the scene in his journal: "Range after range of peaks, snow-clad from base to summit, extended as far as the eye could reach, in splendid confusion."

The confusion was more Barnes's than the mountains'; his party had just spent weeks ascending a tributary divide only to realize they had to descend 3,000 feet back to the main valley of the Elwha. But his words would prove prophetic in more ways than the captain could know. A splendid confusion reigned among geologists banging away at Olympic rocks for the better part of the next century.

Field geology, like all natural science, is a process of building on past knowledge, evaluating new data, reexamining assumptions, questioning earlier hypotheses. The knowledge that geologists have gained to date about the origins of the Olympic Mountains, though extensive, is certainly not the last word — as new dating techniques applied to Olympic rocks have recently shown. Nonetheless, the past 25 years have proved a watershed in our understanding of the geologic story of the Olympics. The plate tectonics revolution of the 1960s cracked open prevailing theories of the Olympics' origins — as it did all of theoretical geology — and the pieces are still being reassembled. Its impact on geology at the time was equal to that of Darwin's theories on 19th-century biological science. Suddenly, the young, rough-hewn rocks of the Olympic Mountains were seen through a new lens, and the deeper story of this rugged range, so recently arrived on the North American continent, began to come to light.

Geologists now know that the Olympic Mountains were formed of sedimentary and volcanic rocks laid down over millions of years on a seafloor off the continental margin. They were later rafted in, crumpled, and plastered onto the edge of the continent like a massive barge running aground on a quiet shoal. But how we've come to know this story, and the painstaking work by which geologists pieced together the scrambled puzzle of the Olympics' orogeny is as fascinating as the sheer heights and shaded canyons of this long-hidden range. That story is a window into the depth and beauty of these mountains as sure as an evening alone on a ridge top, listening for the first star.

Pieces of the Geological Puzzle

Early observers in the Olympics assumed that, like many other mountain ranges, bedrock of the mountainous core of the Olympics must be made up of granite and gneiss. These rocks were commonly found as pebbles and boulders in streams flowing north and east out of the mountains. The Press party explorers brought back a first glimpse of the real story; they correctly identified much of the core rocks as slates and sandstones. But the specimens they had gathered and packed for months over the mountains were lost when their raft capsized in the Quinault River. From such inauspicious beginnings, the story of the origin of the Olympics slowly began to unfold.

In his 1888 report, Governor Semple had described the "continuous array of snow-clad peaks" that bordered Hood Canal to the east and the Strait of Juan de Fuca to the north as composed of basalt, a tough, dark, volcanic rock common to ocean floors. "Pillow" formations in the eastern Olympics suggested that the lava that formed these rocks erupted underwater and cooled quickly. By the 1930s, geologists had identified these basalt rocks as part of the Crescent Formation, a thick horseshoe of basalt that rings the mountains on their north, east, and south sides and is open to the west. Crescent Formation basalts form the bedrock of some of the Olympics' most visible and dramatic peaks: Mount Storm King, Mount Angeles, and Tyler Peak to the north, Mount Constance and the double summits of the Brothers on the east, and Mount Tebo and Colonel Bob to the south.

By piecing together the Crescent basalts with the sedimentary rocks of the interior Olympics and the younger sediments that lay outside the basalt horseshoe, geologists reasoned that the Olympic Mountains were formed from an anticline, a giant fold in the earth's crust that was tilted down toward the continent and whose uppermost rocks had eroded away. This theory neatly explained how the basalt horseshoe of the Crescent Formation formed around what were believed to be the older sedimentary rocks of the interior. Fossils found in the Crescent rocks dated to the Eocene epoch, 38–54 million years ago, but no fossils had been located in the sedimentary core rocks. The anticline theory of the Olympics' formation, the mountains' folding and uplift caused by "tectonic forces deep in the earth's crust," would remain intact for the next 40 years. But throughout that time, the story became increasingly complex, and research brought almost as many questions as answers.

Oil company geologists studying the western Olympics found fossil foraminifera in the sedimentary rocks. These microscopic, chambered

Janis Burger

protozoa were ubiquitous in the earth's oceans over a long period, and they evolved rapidly. This enabled paleontologists to date layers of marine rock with a high degree of accuracy. The age of these fossils showed that some of the core rocks were actually younger than the Crescent basalts. This finding cast doubt on the anticline theory. If the theory was to hold, it would need some serious modification. The plot, as they say, was beginning to thicken.

Mount Constance and the east Olympics are carved from lavas that flowed out on the ocean floor during the Eocene. As the ocean plate collided with the continent, these lavas and associated sedimentary rocks were rafted onto the continent's edge to form the Crescent Formation.

Studies of the Olympics' rocks intensified in the years leading to World War II, when manganese was in great demand. Additional pieces of the puzzle began to fall into place, and an improved geologic map of the Olympics was developed. Even younger fossils were found in the core rocks. They dated from the Oligocene epoch, around 22–38 million years ago. The core rocks were incredibly faulted, or broken up perpendicular to their strata, and restacked, and geologists assumed the basalts were that way too. Scientists now knew what the mountains were made of, how the rocks came to be, and when some of them were deposited. Some of Barnes's splendor was beginning to wear off the confusion. But the larger story of how the mountains were born remained hidden.

The Floor of an Ancient Sea

Throughout Tertiary time, which began about 60 million years ago, the coast of North America lay farther to the east than it does now. The area that was to become the Olympic Peninsula lay quietly beneath the waves of a coastal sea. Rivers pouring off the western edge of the continent carried silt, sand, pebbles, and rocks. Where rivers met salt water they slowed, dropped their load of materials, and formed deltas, just as rivers do today. Rivers, or perhaps a single river — an ancestral Fraser or Columbia — eroded low mountains of the old continent, in an area now roughly comprising western Washington, northern Idaho, and southwestern British Columbia, and vast deltas built up on the continental shelf. As deltas rose higher on the edge of the shelf, storm-related floods or tremors in the earth's crust would trigger an occasional collapse. Undersea landslides of sediments peeled off the deltas and flowed out over the shelf onto deep ocean floors in dense slurries called turbidity currents. Depending on the frequency and nature of the events, some sands settled out in thick beds; others interbedded with layers of finer sediments as silt rained down through calm water between flows. As layers thickened over millions of years in the cold depths of the sea, sedimentary rocks formed. Sands compacted into sandstone, finer silts into mudstones and shale. Occasionally slurries would contain pebbles, forming conglomerate, or broken chips of mud and rock, which became breccia.

As the process repeated itself, regular beds of sandstone and shale formed in layers. Evidence of these rhythmic beds can be found today throughout the Olympics. Particularly noticeable outcrops appear on Blue Mountain along the Deer Park Road. Sandstones, made up of grains of quartz, feldspar, and mica, can be easily distinguished. Shales, while generally composed of the same minerals, are made up of grains much too fine to see. Very little of the sedimentary rocks of the Olympics were buried deep enough or long enough for metamorphism to occur.

Geologists working in the Olympics have all noted the intensity of deformation in the sedimentary core rocks. Instead of neat beds stacked up chronologically like the rocks of the Grand Canyon, strata laid out horizontally along the ocean floor were bent, broken, smashed, smeared, and reshuffled as though kneaded by a very large baker. Rarely do rocks become this deformed without having undergone more extensive metamorphism. The degree of deformation of the Olympic rocks serves to illustrate the tremendous forces that led to the building of these mountains. But before plunging into that part of the story, we'll need to step back a few decades and revisit the state of geological knowledge as it stood in the 1950s.

The Plate Tectonics Revolution

World War II has been described as a technological piñata. The sensitive technologies developed for locating German submarines proved a windfall for oceanographers mapping the ocean floor. In the years that followed World War II, detailed topographic maps of the ocean floor revealed some fascinating features. Long mountain ranges marked by central rift valleys ran lengthwise down the centers of most ocean floors. Seafloors rippled off along either side of the ridges, then dropped into deep trenches that bordered the edges of continents. By towing magnetometers behind ships, geologists were able to measure the intensity of the earth's magnetism as recorded in ocean floors. They discovered some surprising patterns.

Ocean floors are composed of basalt, and basalts are rich in iron. When molten lava cools to form basalt, its iron molecules align with the earth's magnetic field. Scientists have long known that the earth reverses its magnetic field every several hundred thousand years; the last magnetic field reversal was about 720,000 years ago. Cooling basalt retains the magnetic polarity of the earth at the time it solidifies. Scientists working in the north Pacific discovered a striped pattern of magnetic reversals spreading out along either side of the ocean ridges. The patterns were identical to each other. Geologists identified similar magnetic reversals along older lavas on land. Later core dating showed that ocean floors became progressively older the farther one progressed from mid-ocean ridges, with the oldest dates near the edges of continents. But the ocean floor dates in general were curiously young. At 150 million years, the oldest dated seafloors proved much younger than the oldest rocks on the continent, which are 3.8 billion years old. This raised an obvious question: What happened to the older seafloors?

By the mid-1960s, an explanation emerged. It became clear that seafloors are formed along central spreading ridges by lava upwelling from the earth's mantle. As ridges form, new seafloor spreads out along either side, creating plates. These ocean plates interact with other plates also floating on the earth's mantle. Some plates are continents; some part continent, part ocean floor. Some plates override each other; some collide, giving birth to mountain ranges; others split apart to form new oceans. Earthquakes measured along the coastal margins of some continental plates, including the Puget Sound area, indicate that the denser and heavier ocean plates are bending down under the lighter continental plates in a process called subduction. As subducting seafloors descend beneath continental plates, they are metamorphosed into even denser rock. Sediments dragged down with the sinking plates melt and mix with seafloor and mantle to form lava. Some of this lava returns to the surface through faults and conduits in the earth's crust as volcanoes; some remains deep beneath the surface to form batholiths and sills. Volcanic arcs commonly form inland of subduction zones. The Cascade volcanoes bear this relationship to the subducting Juan de Fuca plate, which lies off the Northwest coast. Similar volcanic arcs make up the "Ring of Fire" that circles the Pacific Ocean, delineating the boundaries of the subducting Pacific plate as well as numerous small plates.

An obvious question, and one posed by geologists early in plate tectonics theory, is what drives the plates? The forces that propel seafloors away from ocean ridges are not well understood, though a possible explanation has been offered in the model of convection currents. At the heart of the earth is a molten nickel-iron core. This is surrounded by the semifluid or plastic mantle. Seafloors and continents float on this mantle like the thin film that forms in a pan of heated milk. The earth's mantle may behave somewhat like milk warming in a pan. Just as warming milk rises, pushing cooler, heavier milk off to the sides of the pan and down, convection currents in the mantle may be pushing the ocean plates away from their spreading centers. When these heavier ocean plates collide with the lighter drifting continents, they tend to subduct. But in the area of the Northwest coast that was to become the Olympic Peninsula, the subduction process hit a major snag.

Around 50 million years ago, a chain of basalt seamounts, undersea volcanoes, or a wide lava plateau began to form on the ocean floor.

There's a lively debate among contemporary geologists as to the nature of these basalts and how they came to be. They may have been caused by the ocean plate passing over a hot spot, the way the Hawaiian Islands formed, or by an unusual upwelling of lava through fissures in the ocean crust. Some geologists think the basalts may have been part of a spreading ridge itself. An intriguing new theory accounts for the basalts as the result of rifting and spreading of the seafloor associated with the complex interactions of the Farallon, Kula, and North American plates beginning about 60 million years ago. The rift may have led to massive upwellings of lava along the continental margin. We know by the presence of pillow lavas and marine fossils that virtually all the basalts formed beneath the sea. But rare exposures of columnar basalt in the upper part of the Crescent Formation along Hood Canal suggest that some lavas may have flowed above sea level. Breccia containing large granite boulders found in basalts up the Quilcene River as well as pebble conglomerates around Blue Mountain suggest that the Crescent basalts were formed not far from the continent's edge.

At around the same time, the subducting Farallon plate rafted the basalts and their accumulated sediments into the westward drifting edge of the North American continent. There is some evidence that this movement may have been precipitated or intensified as a result of the Basin and Range expansion to the south, which may have caused a bulge to form in the Farallon plate in the location of the soon-to-form Olympic Mountains. This could account for the the extent of the Olympics' subsequent uplift. At any rate, it's evident from the Leech River fault on southern Vancouver Island to the north, where Cres- | *A cross section of the Olympic Mountains.*

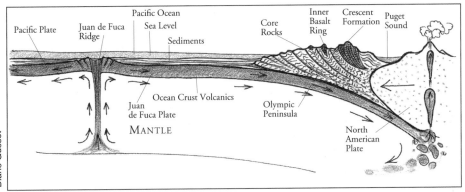

Upended beds of sedimentary rocks alternate with ribs of volcanic breccia on the shoulder of Mount Angeles. The breccias formed as lava flows picked up fragments of existing rock. Mixed beds like these show that Olympic rocks formed close to the continent's edge.

Janis Burger

cent basalts underlie older rocks of the Pacific Rim Terrane, that the basalts began to subduct beneath the continent. But whether it was their size or their shape, the basalts and sediments were too much for the trench to swallow. Instead, the basalts jammed the trench, collided with the North American continent, and partially upended. The collision may have generated enough force to warp the basalts into their present horse-shoe shape as they were crammed into an embayment formed by Vancouver Island and the mainland to the east. It may also have caused the Canadian Gulf Islands to ripple up in a series of linear thrust faults to the north. The younger sedimentary rocks that overlay the basalts were upended and folded as well.

Obviously, many questions remain about the collision and uplift of the Olympic Mountains, but the docking of the basalt and sedimentary formations onto the continent must have been a spectacular event, something akin to seeing today's Olympic Mountains rise up and slam into downtown Seattle. The collision lasted millions of years as the sedimentary rocks that had been laid out on the ocean floor continued to raft in and collide with the basalts, crumpling, shearing, cramming beneath and stacking up behind them like a slow multi-car collision. It's not surprising that one geologist described the whole Olympic range as standing up on end. In retrospect, it's a small wonder that early geologists were able to make any sense out of these rocks at all.

As the Farallon plate continued to subduct, ocean bottom sediments were driven deeper into the earth's crust, where pressure and heat began to chemically alter or slightly metamorphose some of the rocks. Sandstone became semischist, and shale compacted into dense seams of slate and phyllite.

Today, the boundary where the sedimentary core rocks ran up against the upended base of the basalt ridge and were thrust beneath it is marked by the Hurricane Ridge fault. A series of roughly parallel thrust faults echoes through the interior Olympics where the core rocks themselves broke and were thrust beneath each other. Beginning at Cape Flattery in the northwest corner of the peninsula, the Hurricane Ridge fault extends eastward behind Mount Angeles and Blue Mountain before bending south to trace the western wall of the rugged line of east Olympic summits from Buckhorn Mountain south to Mount Washington and Mount Ellinor. From there, a southern extension of the fault bends westward to the Quinault Lake area. On one side of the fault lie the Crescent basalts, on the other the sedimentary rocks of the core.

But as one might expect in the Olympics, the seam isn't quite that clean. With all the folding and faulting accompanying subduction, large sections of basalt were sheared off and placed within the eastern core rocks. This "inner basalt ring" rears up in the jagged spires of the Needles and Sawtooth Ridge. These formations were far more resistant to erosion than the sedimentary rocks around them. Today they offer some of the best rock climbing in the Olympics.

Dates differ, and new research is causing geologists to reconsider earlier ideas, but at some point, possibly as recently as 12 million years ago, the lighter weight of the Olympic rocks relative to the subducting seafloor caused them to bob up like a cork. The radial drainage pattern of the Olympic Peninsula's rivers and recent chemical analysis of the core rocks suggest that the initial uplift of the Olympics may have resembled a massive dome. Almost immediately, as the ancestral Olympics emerged from the sea, the erosive powers of running water and gravity began working on the young rocks.

Carving of the Mountains

The same sea that cooled the lavas and sorted out the sands and silts brought forth the moisture and weather that began to reclaim the rocks grain by grain. Rainfall and melting snow moving across the surface of rocks found weaknesses: faults softened by grinding and pressure, the less resistant siltstones and shales. Aided by gravity and the abrasive power of the silt, sand, and gravel they contained, streams quickly began to carry the mountains away. As the Olympics rose higher and intercepted more moisture from the sea, streams and rivers swelled in size. Larger streams carried more materials, gradients steepened, and the cutting powers of streams increased. The rivers of the western Olympics, the ancestral Hoh, Queets, and Quinault rivers, carving into weaker sedimentary rocks, captured larger watersheds than the streams draining the more resistant basalt bedrock to the east. The Elwha River, which flows north, captured the largest watershed in the Olympics, eventually cutting into the heart of the uplifting mountain mass. But the finish work, the final sculpting of the Olympic peaks and ridges and deep valleys of today, came only with the turn in the earth's climate that ushered in the Ice Age. In the meantime, vast amounts of rock were eroded off the uplifting mountains and carried out to the continental shelf and ocean floor, laying down the grains and strata for ranges of mountains yet to be born.

The Olympic Mountains continue to rise today. Not much, a modest one hundredth of an inch per year. The most recent "pulse" of uplift concluded about 8,000 years ago following the retreat of the continental ice sheet. The Farallon plate that gave birth to the Olympics has almost totally subducted. A remnant of the old plate, renamed the Juan de Fuca

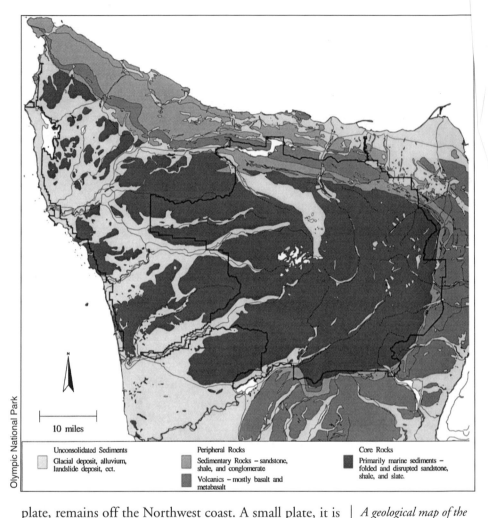

Olympic National Park

Unconsolidated Sediments
Glacial deposit, alluvium,
landslide deposit, ect.

Peripheral Rocks
Sedimentary Rocks – sandstone,
shale, and conglomerate

Volcanics – mostly basalt and
metabasalt

Core Rocks
Primarily marine sediments –
folded and disrupted sandstone,
shale, and slate.

A geological map of the Olympic Peninsula.

plate, remains off the Northwest coast. A small plate, it is presently subducting beneath the continent at a rate of about two inches a year. Earthquake patterns suggest the Juan de Fuca plate might be breaking up as it grinds beneath the continent. But lest we lapse too deeply into complacency, previous large earthquakes in the Puget Sound region and the explosive fury of Mount Saint Helens offer a sobering reminder that Big Things are still going on beneath this restless corner of the earth. And the window the Olympic Mountains offer into the deep tectonic processes of the earth's crust has barely been cracked open.

Reading the Rocks to Hurricane Ridge

The 17-mile drive to Hurricane Ridge provides an easy glimpse into the foundation of the Olympic Mountains and the processes that formed them. The Hurricane Ridge Road leads from the lowlands along the Strait of Juan de Fuca up through the undersea lavas and related sediments of the Crescent Formation and out onto the high windy edge of the mountainous core. From its lower end in Port Angeles, the escarpment of Mount Angeles lifting abruptly out of the upper forests announces the Crescent Formation basalts. As the road gains elevation, it enters a series of tunnels cut though cliffs of pillow basalt. The "pillows" formed as the surface of erupting lava cooled quickly in seawater, giving the rock a lumpy, doughlike texture. Higher up, on the mountain's northeast shoulder, upended flows of volcanic breccia alternate with layers of shale to form a series of fluted chimneys. The chimneys record intermittent lava flows that picked up chunks of sedimentary crust from the seafloor. During lapses

As lava spilled out onto the ocean floor, it cooled rapidly, forming lumps or "pillows." Pillow lavas can be seen throughout the northern, eastern, and southern Olympics.

Janis Burger

between the flows, silt sifted down out of turbid ocean water to form the interbedded shale. The softer shale eroded more quickly, leaving these pronounced fluted chimneys.

Past the tunnels, the Hurricane Ridge Road cuts through a shoulder of Klahhane Ridge where beds of red limestones are interspersed with darker basalts. The limestones formed as microscopic skeletons of plankton and shells of foraminifera rained down over centuries onto the soft muds of quiet ocean floors. Their red color comes from iron oxides in the basalt. Not long before reaching the top of Hurricane Ridge, the road passes the Hurricane Ridge fault, that major juncture between the peripheral rocks of the Crescent Formation and the sedimentary rocks of the mountainous core. Here, the thin-bedded sandstones and mudstones that underlie the Crescent basalts become bent, twisted, sheared, and deformed where the core rocks were thrust against them as the Olympics docked on the continent's edge.

The text of that process — the awesome forces that crumple up mountains and build continents — is written indelibly into bedrock of the Olympic Mountains, scraped clean by glaciers and readily accessible. As we've seen, the Olympics have been a thorny text to decipher, and geologists are just beginning to glean the range's deeper secrets. What they find will not only provide us with a more accurate picture of how the Olympics came to be, we may be given a clear snapshot of that eternal dynamic by which the earth's youthful seafloors and the ancient bones of the continents give birth to rough new lands.

Past the Hurricane Ridge fault, the rocks become increasingly more deformed — and confusing — as you venture into the core. But once the road swings around Hurricane Ridge and the interior Olympics sweep into perspective, the problems of geology give way to the grand view. A chiseled sea of mountains, crest after ragged crest, leaps into view, a more splendid confusion of ridges, canyons, and peaks than a pilgrim could ask for. To the southeast, the bare broken summits of the eastern core are dominated by Mount Anderson in the distance. To the south, the deep, timbered valley of the Elwha River carries the eye into the heart of the mountains where the snow-capped peaks of the Bailey Range stretch away like a high broken wall. Behind them, a slight bit hazy with distance, the summit of Mount Olympus breaks through its perpetual

Janis Burger

The northern Bailey Range from Hurricane Ridge. Core rocks of the interior Olympics are composed primarily of marine sediments. Siltstones, sandstones, and shales were laid down on the ocean floor by rivers beginning about 65 million years ago.

snowfields and glaciers, distant, blue, and implacable.

The mountains seem poised at the crest of a wave. And in a larger sense they are: cast in from the sea, lifted briefly to their wintry heights, and brought down inevitably by the agents of gravity. The Olympic Mountains make up only a small corner of earth, battered by winds and washed with ocean rain. Yet this small, rugged range encloses the beauty of a newly born earth, and like the brushstrokes in a painting by Sesshu or van Gogh, traces the magic by which it was brought to life.

A Long View of an Ephemeral Range

On a cool, drizzly day in August, I visited the Hurricane Ridge fault with geologist Rowland Tabor and a small group of students. Tabor spent 10 years mapping the fine points of Olympic geology for the U.S. Geological Survey, and his book *Geology of Olympic National Park* is the classic guide to understanding Olympic rocks.

We walked the length of the Hurricane Ridge fault, following the near-vertical strata, pecking discreetly at the rock. (In a chapter on the Olympics published in the *Geological Society of America Centennial Field Guide,* Tabor advises his colleagues to keep their rock hammers concealed "to prevent misunderstandings" with park rangers.) At the fault, and beyond it into the core, the rocks were twisted and bent almost beyond recognition. Tabor explained that the intensive deformation of the core rocks was probably the reason these sediments were scraped off onto the continent rather than subducted under the crust. "It all happened in such a short time that the Olympic rocks never got buried deep enough for serious metamorphism to occur," he said. "That's why, unlike the Cascades, there's no gneiss or schist in the Olympics." He explained that the big peaks in the interior, like Mount Olympus and Mount Anderson, were primarily sandstone, the weaker shales and silt-stones having dropped out and eroded away. Because of their sedimentary nature, Tabor refers to the Olympic Mountains as "ephemeral." It was only later that day, as we indulged in a bit of speculation on the long-term future of the Olympics, that I learned what he meant by that.

We were walking back to the car just below Hurricane Ridge. Green, forested slopes swept up into rocky ridge tops around us. I asked Tabor how long it will be before the spreading Juan de Fuca plate propels the Olympic Peninsula onto the continental mainland, where it can take its rightful place alongside its earlier tectonic arrivals, the Northern Cascade and Okanogan subcontinents. He answered with his usual depth of perspective.

"Like all the coast ranges," he explained, "the Olympics are an ephemeral welt. These mountains will erode away and disappear long before they can become part of the geological record of the continent." He looked around him, smiled, then added, "We're lucky to be around to appreciate them."

2
Legacies of Ice

It was late spring, still a bit early in the climbing season for Mount Olympus, and the weather was a mix of low overcast and high wind that climbers refer to as "unstable." Two friends and I had left a chilly camp at Elk Lake and followed a route by flashlight up through Glacier Meadows to the moraine above Blue Glacier. The trail was still deeply buried in winter snow, and a litter of limbs, needles, and lichen blown down from the forest canopy further darkened the route. We reached the moraine just as the overcast sky began to brighten, roped up, strapped on crampons, and began to thread our way up the broad frozen slope of the glacier.

As dawn broke, a wintry landscape of snow and ice fields spread out before us. Dark shapes of cliffs rose into the cloud ceiling, and the occasional *schush* of snow and pebbles whispered down shallow gullies. Light flurries ghosted in on the wind, and the summit rocks remained hidden in cloud.

This was my first visit to the icy realm of Olympus, and though I've returned often in the years that followed, those initial sensations remain crystal clear. As we passed beyond the last shapes of alpine trees, we had the distinct sense of stepping back into an earlier time. That sense deepened as we climbed the crystalline slope that led to the crest of the Snow Dome. The high sloping shelf of the dome spilled its burden of snow in an icefall to the glacier below. As the ceiling lifted, the lower shoulders of the summit rocks drifted into view: snow-dusted, rimed with hoarfrost, and whistling faintly in the mountain wind.

At 7,965 feet in elevation, Mount Olympus and the retinue of peaks that surround it strike visitors as an arctic island within the temperate sea of Olympic's forests. Fueled by relatively warm, moisture-laden winds from the Pacific Ocean a mere 35 miles away, Mount Olympus receives some 240 inches of precipitation each year, the highest amount in the lower 48 states. It falls mostly in the form of snow, over a hundred feet in the course of a winter. The snowpack feeds a prodigious system of glaciers on Olympus; nine alpine glaciers descend from the upper snow slopes into the headwater valleys of the Hoh and Queets rivers. Given the amount of snow that falls on these mountains, it's not surprising that the glaciers of Mount Olympus form at a lower elevation and descend farther down into their valleys than anywhere else in the lower 48 states. Blue Glacier is the largest of the Olympics' 266 glaciers; it descends two and three-quarters miles to as low as 4,049 feet in elevation, where it pours its opaque, glacier flour-laden stream into the headwaters of the Hoh River.

The Blue Glacier is a stunning contemporary example of the shape and dynamics of a force that thoroughly transformed not only these mountains but much of the northern landscape of North America. It is also one of the most intensively studied glaciers in the country.

In geologic terms, glaciers are a recent phenomenon in the Olympics, appearing only over the past 1.8 million years. When considering them, we need to hone our sense of geologic time — from thinking in terms of millions of years to thousands. In the case of the Little Ice Age, we can further reduce that time scale to mere centuries — barely the dust settling on the open page of a book.

In the relatively short geological lifespan of the Olympics, the recent 1.8 million years of the Pleistocene epoch was a brief interval, the shadow of its passing still upon us. But its effects on the rolling meadowlands and smooth, water-carved preglacial Olympics was profound. Had it not occurred, the Olympic Mountains might resemble the gentle folded hills of the Coast Range to the south. And the unique biological diversity so characteristic of these mountains would surely never have developed.

However recent and brief their tenure here, glaciers marshaled incredible force in shaping the face of the Olympic Mountains as we know them today. From serrated ridges and steep mountain walls to rounded foothills and broad valley floors, evidence of the Olympics' glacial legacy is carved irrevocably into the breadth and texture of the land. Not only did alpine glaciers sculpt the present shape and character of the landscape, large ice sheets advancing from the north wielded immense power in sculpting and forming the character of the entire region. That the Olympics form a peninsula today, wreathed by seawater and isolated by a wide valley to the south, was largely the work of the Cordilleran ice sheet pulsing south from its chill northerly realm. And with all the grace of a hundred thousand D-8 Cats with an ocean of diesel to burn, it changed things around some.

Dawn of the Pleistocene

The forces that ushered in the glacial epoch are complex, involving grand-scale interactions among continents, oceans, the atmosphere, and the solar system itself. It's evident from the geological record elsewhere that there have been successions of glacial ages in the past, as far back as 2.3 billion years ago, each lasting tens of millions of years. Climatologists suspect that these events were triggered by a combination of the earth's tectonic forces (movements of continental plates, collision and uplift, and the birth of oceans), and their resultant effects on atmospheric circulation and ocean currents. Scientists generally agree that for an ice age to occur, much of the earth's continental land mass must be located in the higher latitudes, a condition certainly present in the late Cenozoic, which preceded the Pleistocene. As plates collide and ranges are uplifted, as they were in western North America and central Asia during the Cenozoic, continental crust is exposed above the level of soil and plants, and continents reflect more of the sun's warmth, reducing the amount of solar energy available at critical northern latitudes. Higher ground also collects more snow. It's also clear that reduced summer melting rather than increased winter accumulation of snow is key to the formation of ice caps from which continental glaciers are born. Once perpetual ice becomes established it reflects more of the sun's warmth away from the earth, and a ripple effect spreads through the atmosphere and ocean currents, leading to feedback loops of cloudier skies, less warmth — and more ice.

Once a glacial age begins, cyclical variations or glaciations within it are triggered by subtler factors involving the earth's pattern of rotation and tilt and the solar cycles of the sun. We also know that two critical "greenhouse gases," carbon dioxide and methane, were less plentiful in the atmosphere during glacial times, allowing radiant energy from the earth's surface to escape into space, but it's unclear why. The ice ages that spawned the glaciers that changed the face of the Olympics began with a gradual cooling of the earth's climate around two and a half million years ago.

Whether an alpine glacier flowing into a valley or a continental ice sheet grinding down from the north, glaciers are formed and behave in essentially the same way. Simply stated, glaciers form in regions where winter snowfall exceeds summer melt. In the Olympics today the snowline (the elevation where snow lingers from year to year) occurs at a relatively low 5,500 feet. By contrast, snowline in the northern Rockies, with that region's warmer summers and lower precipitation, lies at around 10,000 feet. Closer to home, snowline in the Cascade Mountains occurs at about 8,000 feet. Above snowline, the process whereby winter snow compacts into glacial ice is gradual but complete.

During the warming days of spring, individual snowflakes are melted, compacted, and recrystallized into rounded granules or corn snow. As compaction increases, a hard permeable crust of firn snow develops. Under the weight and pressure of continued compaction, air is driven from the firn and snow granules melt at point of contact, meltwater then refreezing between them. The interwoven, crystalline ice grains that result become further deformed as a glacier becomes "plastic" and begins, under the pressure of weight and gravity, to flow downhill.

This formative process, the birthing of glacier ice, requires a year and at least a 100- to 200-foot accumulation of snow. It also mirrors, in microcosm and in a radically telescoped time scale, the formation of sedimentary rocks. As geologist John Shelton has poetically noted, snow is an eolian sediment. Deposited by wind, layered and compacted, firn snow resembles sedimentary rock. Glacier ice, having recrystallized under tremendous pressure and stress, shares similarities with metamorphic rock. Deep ice melting against substrate mimics deeply buried rock strata melding with the earth's mantle.

Today, these big, soft, blue-white rocks abound on Mount Olympus, where annual snow accumulation more than accommodates them, as well as on many other high summits in the interior Olympic Mountains. Glaciers fare less well in the dryer eastern Olympics, but this wasn't always the case. The scene that greeted my friends and me that chill morning on Mount Olympus was fairly common throughout these mountains on several occasions throughout the Pleistocene, and evidence of glacial carving can be seen everywhere throughout the range. Few are the Olympic peaks whose symmetry and lines do not reflect the sculpting hand of glacial ice. And fewer still are the river valleys left unshaped by glaciers moving their slow trains of rock-bound ice into the lowlands.

Final Sculpting of the Olympics

The Olympic Mountains were affected not only by repeated advances of alpine glaciers but by the Cordilleran ice sheet, which advanced south into the area from the Coast Mountains of northern British Columbia at least six times over the last two million years. The tracks of alpine glaciers are a bit more difficult to read, as successive glacial advances tend to erase the evidence of earlier glaciations. Geologists studying the moraines of alpine glaciers on the west side of the Olympics have identified four major alpine advances, but core samples from ocean sediments indicate there were more than 20 periods of increased glaciation throughout the Pleistocene. It was during this time that the final sculpting of the Olympics took place.

As with most mountain ranges, the alpine glaciers that carved the Olympics tended to follow drainage patterns already incised by rivers and streams. But where the gentler hands of flowing water tended to bend, meander, and flow around resistant strata or formations of rock, the freight trains of glaciers bowled right through. As a result, alpine (now valley) glaciers tended to widen and straighten preexisting stream-cut valleys in the Olympics, shaving back spur ridges and flattening river bottoms into characteristic U-shaped valley floors. This is most evident today in the large western Olympic river valleys such as the Hoh, Queets, and Quinault. In contrast, stream-cut valleys have a marked V shape. As glaciers flowed down the Olympics' valleys, rocks as well as river cobble, gravels, and sand become frozen into the bottoms of glaciers and acted like huge rasps scouring the bedrock. Large boulders and flakes of rock also froze into the glaciers and were plucked and carried along.

When summer melting exceeded winter accumulation, valley glaciers receded, leaving behind their load of rock and debris. Piles of talus and rockfall that accumulated on the tops of glaciers sloughed off their sides to become lateral moraines. Recent lateral moraines are prominent along the northeast side of Blue Glacier and in several deglaciated alpine basins such as those at the head of Royal and Cameron creeks in the Dungeness watershed. The cross-valley piles of rock dumped off the stationary or retreating terminus (or snout) of a glacier are terminal moraines. Quinault Lake in the southwest corner of Olympic National Park formed behind a terminal moraine of the Quinault Valley glacier.

Tremendous amounts of rock and rubble were transported from mountains to valley floors by Olympic glaciers. Materials sorted into rough layers of sand, gravel, and boulders by glacier meltwaters are referred to as outwash; unsorted materials are called drift. Both are easily seen along the cutbanks of lower rivers or in storm-cut sea cliffs along the Olympic coast. Outwash plains resemble alluvial fans often found at the mouths of steep mountain canyons, though the vast majority of ice-age outwash plains in the Olympics have been reshaped by rivers and are now well hidden beneath valley forests. Nonetheless, most of what we have come to know of the behavior of valley glaciers in the Olympics has been pieced together by studying moraines, outwash plains, and sea cliffs.

Glaciers were also the principal agents in shaping the character of the rugged mountainous uplands of the Olympics. Unlike the headwaters of nonglaciated streams which fold themselves gradually, tributary by tributary, into a lower valley, the headwaters of Olympic valleys often end abruptly in steep-walled cirques. These high mountain basins, carpeted with meadows and scattered copses of subalpine trees, often cradling a still mountain tarn at their base, comprise some of the most stunning alpine scenery in the Olympics. The ice-free cirque basins of upper Royal

Creek, Grand Creek, and the Sol Duc River are among the most popular backcountry destinations in Olympic National Park.

Cirque walls are rough-carved and cliffy. As firn and ice along the upper end of a glacier melt away from a rock headwall, a deep *bergschrund* forms. In late summer such "schrunds" become the bane of climbers, often requiring dicey acrobatic passages from glacier to summit rock. As meltwater freezes in cracks behind exposed blocks of rock in the headwall, the blocks loosen and clatter down into the *bergschrund* to become part of the glacier's load. These angular blocks increase a glacier's cutting power and help it grind down the floor of a cirque, further steepening the cirque walls. The cutting power of upper glaciers is also increased by loads from snow cornices collapsing from wind-blown ridges above, or icefalls spilling onto it from an adjoining peak. There's no doubt a cirque of magnificent proportions being carved at the head of Blue Glacier today.

While alpine glaciers grind away at cliff sides, gravity and frost action whittle the upper ridges, sharpening ridge crests, steepening relief, and generally sprucing up the neighborhood. Cameron Basin, one of the most beautiful

Mount Ferry Basin in the Bailey Range present a classic glacier-carved landscape. Ice-sharpened ridges give way to a broad glacier-carved cirque floor and snowmelt tarn.

Janis Burger

subalpine parklands in the Olympics, presents a stunning example of a glacier-carved mountainscape. A large cirque broken by glacier-smoothed cliffs and ridges and veined with gently undulating moraines, the basin is a classic glacial landscape. A hike to Cameron Pass at its head reveals the marked contrast between the steep, glacier-cut basin to the north and the gentle, nonglaciated slope falling away to Lost Basin to the south.

Valley glaciers descended all of the major river valleys in the Olympics, carving wide U-shaped troughs in the sedimentary core rocks and narrower slots in the more resistant basalts of the Crescent Formation to the north, east, and south. Radiocarbon dating of a peat bog below a terminal moraine on the Hoh-Bogachiel divide suggests that the maximum advance of alpine glaciers in the Olympics occurred between 30,000 and 60,000 years ago, well ahead of the most recent advance of the Cordilleran ice sheet into the Olympics. Moraines indicate that the Quinault glacier escaped the confines of its valley walls at this time and may have coalesced with the Queets glacier at one point, forming a small piedmont glacier on the southern coastal plain. The maximum advances of the Hoh glacier and the Dosewallips glacier to the east most likely reached present-day sea level; it is unlikely that any of the Olympic glaciers reached the much lower sea level of that time. Glacial till (drift that hasn't been reworked by meltwater) is present in sea cliffs south of the Hoh River, suggesting that the terminus of the Hoh Valley glacier reached at least that far.

Though difficult to read, sea cliffs reveal a remarkable record of the past. For example, sea cliffs south of the Hoh River show lower outwash strata that seem to be tilted more than more recent strata above them. Geologists interpret this to mean that uplift of the Olympics continued well into the glacial era. Similar evidence can also be found along older river terraces, and these are presently the subject of intensive study.

While terminal moraines of valley glaciers can be found in the western Olympics, evidence of the maximum advance of east-side valley glaciers is harder to find. This is due to the singular fact that all of the northern and eastern lowlands were completely buried and dramatically reshaped beneath thousands of feet of Cordilleran ice.

The Coming of Continental Ice

Although it was almost an afterthought, a final gloss on the geological history of these mountains, no event would have as profound an effect on the ecology of the Olympics as the coming of the Cordilleran ice sheet. At least six times during the 1.8 million years of the Pleistocene, a vast ice sheet descended from western British Columbia into the neighborhood of the Olympics. On at least four of those glaciations, the ice sheet surrounded the Olympic Mountains on two sides. With outwash streams surging around the southern end of the peninsula, the Olympic Mountains were effectively isolated from the nearby Cascades and Coast Range mountains at least four times during their history. This isolation, combined with the topography and geographical isolation of this islandlike range, was a critical factor in determining the patterns of life forms that presently grace the Olympics.

Like alpine glaciers in the Olympics, the Cordilleran ice sheet was also born in snowy mountain cirques, only these were far to the north in the Coast Mountains of British Columbia. Ranging between 10,000 and 13,000 feet, the Coast Mountains are much higher than the Olympics and present a formidable barrier to moist air flowing in off the Pacific. Even today the Coast Mountains support sizable ice caps. Alpine glaciers flowing west from these mountains coalesced into a piedmont glacier in the coastal lowlands. While the northern end of the ice sheet flowed west into the Pacific, its southern expanse ran up against the Insular Mountains of Vancouver Island. Picking up additional mass from the glaciers flowing east from the island mountains, the ice sheet began to flow south.

The earliest identifiable advance of the Cordilleran ice sheet into the area of the Olympic Mountains, the Orting glaciation, occurred about 2 million years ago. When it reached the peninsula, the ice sheet split against the basalt prow of the northeast Olympics. Like the wakes of a bow wave, one lobe flowed west, initiating the carving of the waterway that would be known as the Strait of Juan de Fuca, and one lobe flowed south along the eastern Olympic front into the Puget trough.

As world climate fluctuated throughout the Pleistocene, the Orting glaciation was followed by three other advances of the ice sheet: the Stuck, the Salmon Springs, and, most recently, the Fraser glaciations.

Each of these were in turn marked by pulses or "stades" as the ice pushed forward and melted back in response to climatic conditions. Glaciologists are still sorting out dates and areas affected by various glaciations; the date of the Salmon Springs glaciation, for example, was just bumped back a half million years. Most easily read, however, and evident throughout the northern and eastern Olympics, is the most recent glacial advance into this area, the Vashon stade of the Fraser glaciation.

Beginning a mere 15,000 years ago, the Vashon was one of the most extensive glaciations to affect the Olympic Peninsula. Like the Orting and earlier glaciations, it advanced in two lobes. Hemmed by the northern scarp of the Olympics, the Juan de Fuca lobe ground west, curving around the northwestern end of the mountains to the Quillayute Prairie north of the present town of Forks, and completing the carving of the Strait of Juan de Fuca. The Puget lobe flowed south along the east face of the Olympics. Advancing between 200 and 500 feet per year, it reached as far as the Black Hills south of Olympia and left in its wake the lovely inland sea of Puget Sound. Swollen by the melting valley glaciers of the east Olympics and west slope of the Cascades, the vast braided outwash stream of the Puget lobe cut southwest around the Black Hills and out to the Pacific along the present course of the Chehalis River, leaving a channel vastly oversized for the modest stream that occupies it today. The Cordilleran sheet had reached this far at least once before, maybe more, but each successive advance tended to work over the earlier tracks pretty thoroughly. In the Olympic Mountains, however, evidence of this most recent passage of the Cordilleran is clear to even the most casual viewer.

Blocks and boulders of telltale light-colored granite lie scattered like Easter eggs in river beds and on forested mountain slopes up to nearly 4,000 feet in the northeast Olympics. Since granite is not found in the bedrock here, all of these "erratics" must have been carried south from the mountains of western Canada by the glacier and dropped off as the ice melted. The ice sheet reached its highest point at 3,800 feet on Blue Mountain and tapered off in height as it split off west and south. The Puget lobe was about 3,200 feet thick over Seattle, 1,500 feet over Olympia, and erratics are plentiful throughout the middle and lower elevations.

It's hard to envision what it must have been like in the Olympic

Mountains during the Vashon glaciation. The Malaspenia Glacier that flows out of the Saint Elias Mountains in Alaska today probably provides the closest analogy, but climate and conditions here were far different. The alpine glaciers of the Olympics had reached their maximum advance during the earlier Evans Creek stade of the Fraser glaciation between 22,000 and 18,000 years ago. By the time of the Vashon stade they had retreated part way up their valleys, leaving a drift of bouldery cobbles and outwash. As the Vashon ice sheet dammed the northern and eastern river drainages, valley lakes were impounded and filled with runoff from the receding valley glaciers. Large fjordlike lakes occupied the Elwha, Dungeness, Dosewallips, Duckabush, Hamma Hamma, and Skokomish river valleys. As chunks of Cordilleran ice calved off into the lakes, erratic rocks frozen into the ice were rafted on floes and dropped far up valleys. Overflow streams cut new channels along the margins of the ice sheet, and fine, clay-forming silt settled over lake bottoms. Numerous unstable valley slopes, slides, and road failures in the Olympics today hearken back to these flooded valleys.

Extent of the Fraser glaciation on the Olympic Peninsula showing the estimated extent of alpine glaciers during the Evans Creek stade (around 19,000 years ago) and the extent of the Cordilleran ice sheet during the Vashon stade (around 15,000 years ago).

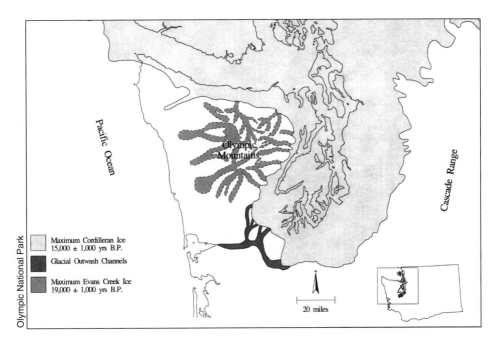

The Vashon ice sheet may have enclosed the northern Olympics for 1,500 years, reaching its maximum extent some 14,000 years ago. As the lobes of ice retreated, glacial lakes occupied the low basins carved by the flowing ice sheet. The Juan de Fuca lobe melted back first, allowing the impounded lakes to drain north and Olympic rivers to resume their preglacial courses. By 12,500 years ago, the ice had left the strait and Puget Sound; the world's climate ameliorated and marine waters filled the troughs left in the ice sheet's wake. Vegetation recolonized the denuded glacial slopes, and large mammals such as mastodon — as well as their two-legged hunters — ventured into the northern lowlands. But the signature of ice was indelibly inscribed in a vastly altered landscape.

Fog fills the Elwha and Lillian valleys, a common occurrence in winter in the Olympics. The view suggests the Pleistocene landscape of 18,000 years ago when alpine glaciers flowed down into river valleys.

Rounded foothills and spur ridges, kettle lakes and scattered wetlands and bogs attest to the heavy hand of the ice sheet. Puget Sound, the Strait of Juan de Fuca, and the long glacier-carved fjord known as Hood Canal mark its track, as do some of the park's most pristine lowland lakes such as Crescent and Ozette. If glacier-bared ground is difficult to envision beneath the lush mantle of lowland forest, it doesn't take much digging with a shovel to bring the past alive. The ubiquitous presence of glacial till

Janis Burger

(known locally as hardpan), gravels, moraines, and rocky drift that underlie the lowland landscape soon make themselves known. For the observant, it's possible to read the geological past by marking the kinds of vegetation, the forest and plant associations that thrive in this area today.

Adaptations of Ice Age Plants and Animals

That glaciers linger in the temperate Olympic Mountains should be evidence enough that we live in a time of climatic change. Whether remnants of an age recently passed or harbingers of a time to come, the sight of living glaciers touches something deep in the human imagination. Cold seems to have more bite in the blue-white wilderness of ice, and the steady drone of the wind more resonance. When midday sun thaws meltwater rills along a glacier's edge, the tinkling song is sometimes met with a half-uttered sigh of reassurance from alpine travelers.

Scientists are curious about the profound climatic upheavals of the past, and how they may help us understand the interglacial age we live in. One way of approaching the big picture also casts light onto current biotic communities of the Olympics: studying the ways in which local plants and animals responded to the fluctuations of Pleistocene and more recent Holocene climates.

The record for Pleistocene animal life in our area is sketchy; glaciers did a pretty good job of smudging out most traces of Ice Age fauna. Marine fossils from this time suggest a colder and generally more northern marine environment than is found here today. Remains of land-dwelling animals indicate that bison, elk, caribou, woolly mammoth, and mastodon found their way into adjoining lowland areas during interglacial periods. The most complete picture of early postglacial animal life on the peninsula came in 1977, with a remarkable discovery about six miles south of Sequim.

Emanuel Manis was digging a duck pond in a peat bog on his land when he turned up two "logs" that resembled tusks. That is what they were; the site eventually yielded the nearly complete skeleton of a mastodon *(Mammut americanum)* estimated to have died about 12,000 years ago. More remarkable than the skeleton itself were signs that the mastodon may have been butchered for meat by humans. The site also showed signs of human encampment and tool use. The Manis site, studied intensively by archaeologists, presents an intriguing view of human and animal interactions as well as the condition of the natural environment on the northeast Olympic Peninsula in the centuries that followed the retreat of ice sheet.

Scientists studying the plants that lived during the various stages of the glacial age have turned up a much more continuous and complete record. By examining pollens found in the sediments of glacial peat bogs and interglacial (and postglacial) lake beds, as well as identifying ash layers from known volcanic eruptions and radiocarbon-dating plant remains, a fascinating picture has emerged of the Olympic Peninsula and Puget Sound area over the past 20,000 years.

Just before the most recent Vashon glaciation, the Laurentide ice sheet had advanced to its farthest extent over the northern third of North America, and its presence caused a general cooling in the Pacific Northwest. The Laurentide also split the jet stream, shifting winter storm tracks to the south and reducing precipitation throughout the Northwest. Cold dry easterly winds from the great ice sheet further deepened the cool and arid conditions here.

As the Cordilleran ice sheet began its advance and alpine and valley glaciers were well into their retreat from the Olympic lowlands, the western Olympic Peninsula was covered by a mosaic of parklands, muskeg and sedge communities, and subalpine forest reflecting a moderating marine influence. Scattered forests of spruce, pine, mountain hemlock, and alder suggested forest communities found at higher elevations in the Olympics today. Areas closer to the ice sheet were open, supporting grasses, sagebrush, and alpine plants. Lowlands on the east side of the Olympics were similarly open and supported such continental tree species as Engelmann spruce — which today is found mostly in mountains east of the Cascade crest as well as in scattered sites in the Olympics — lodgepole pine, and subalpine fir. Grasses, sedges, and herbs dominated open ground in the eastern lowlands, creating an environment somewhat resembling the forests of the northern Rockies.

As the Cordilleran sheet reached its maximum extent, around 15,000 years ago, the climate had already begun to warm. More familiar lowland forest species such as Sitka spruce and western hemlock began to appear in the coastal Olympics. Conditions probably became wetter during this time, because of the recession of the Laurentide ice sheet and a return to "normal" winter storm patterns. Temperatures were probably three and a half to eleven degrees Fahrenheit colder than today's.

As the Puget and Juan de Fuca lobes receded north from their moraines and outwash plains around 14,000 years ago, the first tree species to colonize the rocky outwash soils was lodgepole pine. The pollen record shows that it was 2,000 to 4,000 years before the pines were joined by Sitka spruce, Douglas-fir, western hemlock, and alders to form a closed-canopy forest in these areas. In the arid rainshadow of the northeastern Olympics, open communities and herbs and shrubs persisted until about 10,000 years ago, although prairie and oak woodlands had colonized the coarser outwash soils as early as 12,000 years ago. Pollen from core samples taken in southern Puget Sound, beyond the margin of the ice sheet, show that high-elevation species and lowland forest species commingled following the glacier's retreat between 10,000 and 11,000 years ago.

The warming trend in the earth's climate that led to the rapid retreat of the ice sheets and the subsequent rising of sea levels continued for several thousand years, reaching a time of maximum warming between 9,000 and 10,000 years ago called the Hypsithermal period or climatic optimum. An increase in the tilt in the earth's axis and its position relative to the sun at this time intensified the solar radiation reaching the earth. This led to near-Mediterranean conditions of higher temperatures, lower precipitation, and extreme summer drought that prevailed until about 6,000 years ago. Evidence of the Hypsithermal warming is evident in the pollen record on the Olympic Peninsula as fire-responsive species like Douglas-fir, red alder, and bracken fern became more abundant within the older, moisture-loving spruce-hemlock forests. As wildfire played an increasingly important role in the ecology of the area, prairies expanded in the lowlands, and plant communities of the Puget basin began to resemble the open savannas of the Willamette Valley in Oregon. It was during this time that trees more common to southern climates, like the lovely Pacific madrona and golden chinquapin, may have found their way to the Olympics, adding to the beauty of our lowland forests.

Sometime between 5,000 and 6,000 years ago, the intensity of summer droughts began to abate and the "grandfather tree" of Northwest coastal people, the western redcedar, became plentiful along streams and coastal lowlands. The return of redcedars was a momentous event for the Olympic Peninsula. Not only would this tree provide the material foundation for a highly developed Native American culture, it prepared the way for the return of salmon to peninsula rivers by falling into streams and creating pools and rifles. The redcedars were soon followed by western hemlock, Sitka spruce, and western white pine, a return to forest communities similar to those found on the peninsula today.

Not long ago, it was assumed by many ecologists that forest and plant communities, along with their associated wildlife, "migrated" north and south along elevation gradients and mountain ranges with the advance and retreat of glaciers. Appealing as this idea may be, careful analysis of the pollen record shows that plants responded to climatic change and disturbance in subtler and more complex ways. For instance, during the time of maximum Cordilleran advance, temperate plants were thriving close by and trees were present on nearby uplands. Indeed, the quick return of Douglas-fir to the peninsula and nearby areas during the Hypsithermal (less than 500 years after conditions grew warmer) and the rapidity with which other forest trees responded to late glacial changes suggest that complex adaptations were at work. To understand these processes, to explain some of the wonderful diversity of the Olympics' plant communities, and to explore the significant role the Olympics played in preserving biodiversity in the Northwest forest ecosystem, let's look into the idea of biological refuges.

The Olympics as Biological Refuge

Ecologists call them *refugia*, places where plants and animals can escape extreme conditions. They can be cool bogs or upland areas in hot conditions, lakeshores and wetlands during arid times, or ice-free places during periods of glaciation. As we've seen, just in the past 20,000 years the Pacific Northwest experienced a wide range of climatic extremes including the prolonged visits of an Arctic ice mass. It was a period in which no millennium was exactly like any other. For a *refugium* to harbor a variety of plants and animals through these changing conditions — as well as throughout the 1.8 million years of the Pleistocene — it would have to contain an incredible diversity of habitats, microclimates, elevational ranges. It would have to be unique.

To Nelsa Buckingham, a botanical researcher on the Olympic Peninsula, the explanation is obvious, and supporting evidence is everywhere in the kinds of plants and animals that are found — and not found — on the Olympic Peninsula today. Buckingham, along with a growing number of scientists in a number of fields, believes that the Olympic Mountains served as such a *refugium* for plants,

Mean precipitation in inches for the Olympic Peninsula. Differences in precipitation create an abundance of habitat niches in a relatively small area.

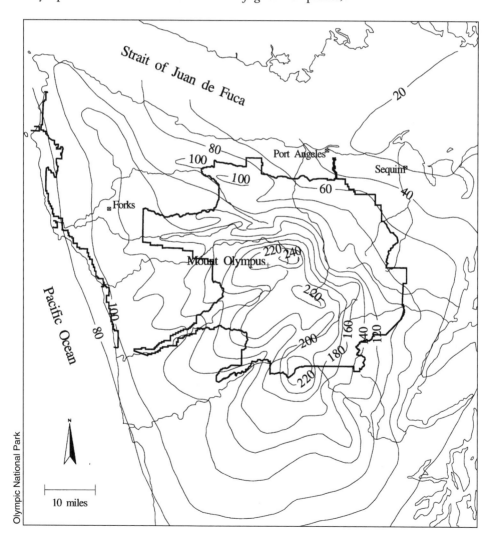

not only through the ice ages of the Pleistocene, but as far back as the late Miocene, 5 million to 15 million years ago. To illustrate, Buckingham points to a number of biological characteristics that make the Olympic Peninsula unique. Though the peninsula shares most of its plant species with other areas of western Washington, it lacks some species, such as noble fir, that are common elsewhere in the state. The peninsula also hosts a surprising number of endemic plant species, plants that occur here and nowhere else. There were eight of these endemics at last count. Another six occur only here and on Vancouver Island to the north, and three more are found only here and on Saddle Mountain on the Oregon coast. Just as fascinating to Buckingham are the peninsula's "disjunct" plant species, those found here as well as some distance away but missing from areas in between.

Other aspects of the story intrigue Buckingham as well. Though the peninsula occupies only 9 percent of Washington State, it harbors 27 percent of the plants listed as rare by the Washington Natural Heritage Program. Add to that the nine endemic animal species and subspecies found in the Olympics (five mammals, one amphibian, three fish), as well the 12 mammal species common to the Cascade Mountains but historically *absent* from the Olympics, and it becomes obvious that some significant evolutionary factors have been at work here.

A number of factors contribute to this unique situation. Topographically, the abrupt relief of the mountains (less than 10 miles from tidewater to the alpine summit of Mount Angeles) has created a range of elevational life zones in close proximity. Climatically, the same moisture-laden southwest winds that dump 240 inches of precipitation on Mount Olympus every year are wrung dry by the mountains, sprinkling a mere 18 inches on the rainshadow area of Sequim less than 34 miles away and creating one of the steepest precipitation declines in the world. This combination of climatic and topographic diversity yields a wealth of habitats and niches. During times of severe climatic change, alternative habitat sites for plants as well as animals are likely to be close by.

Perhaps the most significant factor leading to the uniqueness of the peninsula's biotic communities is topographical isolation. The Olympics are an islandlike cluster of mountains separated from neighboring ranges by a lowland river valley and inland sea. The peninsula was also isolated periodically by continental ice sheets and outwash channels during the long millennia of the glacial age. This isolation tended to limit the ranges of a number of peninsula plants, possibly allowing distinct subspecies and varieties to evolve. It also hindered the repopulation of a number of wildlife species, particularly those that prefer mountain habitats, after the ice had receded.

As far back as the Miocene, the Olympics shared the subtropical vegetation that stretched from the Pacific Coast east to the Rockies. The uplift of the Cascades between 2 million and 7 million years ago blocked the easterly flow of moist marine air and gave rise to vast arid prairies east of the Cascades. Olympic Peninsula plants were further isolated by the repeated advance of the ice sheet over the past 2 million years as the ranges of many plants to the north and east were erased by the ice. Since, at the height of glaciation, only the northern and eastern margins of the peninsula and the glaciated upper valleys were covered by ice, most of the landscape, including a large area between the Dungeness and the Elwha rivers — and a wide coastal plain exposed by the lower sea level (as much as 400 feet below current levels) — remained an ice-free refuge for plants.

To illustrate this for me, Olympic National Park botanist Ed Schreiner called up some range maps for Olympic plant species on the computer screen in his office. Based largely on the work of Buckingham and others, base maps of the North American continent showed the disjunct ranges of plant species like sticky oxytropis (*Oxytropis viscida*). It occurs on the peninsula, but its next closest range is in the mountains of northeastern Oregon. Another disjunct is long-stalked draba (*Draba longipes*), whose nearest kin lies in northeastern British Columbia, more than 300

Endemic Plants of the Olympic Peninsula

Olympic Mountain Milkvetch (*Astragalus australis var. olympicus*)
Piper's Bellflower (*Campanula piperi*)
Flett's Fleabane (*Erigeron flettii*)
Thompson's Wandering Fleabane (*Erigeron peregrinus ssp. peregrinus var. thompsonii*)
Olympic Mountain Rockmat (*Petrophylum hendersonii*)
Olympic Mountain Groundsel (*Senecio neowebsteri*)
Olympic Mountain Synthyris (*Synthyris pinnatifida var. lanuginosa*)
Flett's Violet (*Viola flettii*)

miles distant. The northern grass-of-Parnassus *(Parnassia palustris)* has a similar range; it is found here and in the northern interiors of northwest Canada and Alaska. Mountain owl clover *(Orthocarpus imbricatus)* graces meadows in the Olympics and on Vancouver Island but isn't found again until one travels to the mountains of southern Oregon. Many plants, including Cooley's buttercup *(Kumlienia cooleyae),* appear to be at the southern extent of their ranges along the Pacific Coast in the Olympics. It's logical to suspect that the ranges of all these species were once joined to the Olympic Peninsula but that tectonic events, climate change, and the movements of Pleistocene ice separated them. Following the glaciers, the spread of closed-canopy forests in the lowlands may have further isolated these and many other disjunct species.

Schreiner thinks that the ranges of these and other disjuncts present a pretty clear case for the Olympics as an ice-age refuge. He compares the Olympic Peninsula to the "land bridge" islands to the north, the Alexander and Queen Charlotte archipelagos and Vancouver Island. All were connected to the mainland during parts of the Pleistocene, and all are now isolated. Endemic and disjunct plants as well as "missing" animal species are characteristic of islands, land bridge islands, and island habitats like the Olympics worldwide, but it's the particular assemblage of plants and animals found here, says Schreiner, that stamps the peninsula as unique.

Questions certainly persist. Did the endemic plant species of the Olympics — plants such as the stunning Piper's bellflower and delicate Flett's violet — evolve here, or are they the relicts of once wider-spread populations? (Schreiner suspects that the Olympics have both.) How did — and how do — the Olympics interact with other land bridge islands farther north? One thing is certain: the ice ages and the climatic upheavals that preceded them have reshuffled the deck of plant and animal communities wildly in the Olympic Mountains. It's a common characteristic of glacial *refugia* that plant species tend be all mixed up — alpine and subalpine species mixing willy-nilly with lowland plants and varieties. When climatic shifts occur on the order of centuries, species get by where they can. Nelsa Buckingham likes to point out that 75 percent of the nearly 1,200 vascular plant species and subspecies native to the Olympics occur from the lowlands to the subalpine, and those that don't are usually confined to specialized habitats like lakes and shorelines. Her conclusion is that not only have the Olympics served as a *refugium* since the Miocene, these mountains remain a refuge today.

Certainly the Olympics hold a key position in the region's evolutionary history, and they may prove just as vital to its future. With the prospect of global warming of seven to nine degrees Fahrenheit facing us in the next century, diverse habitats like those of the Olympic Mountains and the species richness they harbor could prove essential to the continuing process of evolution.

The Little Ice Age

When we view the climatic fluctuations of the past, we tend to do so from the supposed "norm" of our current climate. The recent decades of our interglacial age have given us a mistaken sense of climatic stability; the very phrase "ice ages" summons images from a primordial past. But as recently as a few centuries ago, world climate was in the midst of a tenuous half-step back toward that earlier time of ice. The evidence of this global cooling, along with its very recent aftermath, was recorded quite clearly by the alpine glaciers of the Olympics.

Since the Neoglacial or Little Ice Age occurred during the historic era in Europe, its evidence — both direct and indirect — abounds in the historic record. Lithographs and paintings dating to the 16th and 17th centuries show glaciers lying close upon alpine villages in Europe; historic documents confirm that the villages of LaRoziere and Argentier in France were overrun. Wheat prices in Europe escalated between 1250 and 1350, then again in the 1600s and 1800s, correlating with cycles of cold weather. In the 1400s, Vikings found sailing routes to their Greenland colonies blocked by ice. Closer to home, Captain Vancouver, sailing the Northwest coast in the late 1700s, recorded that Alaska's Glacier Bay was filled with ice.

Alpine glaciers extended worldwide, and the glaciers of the Olympics were no exception. Though nearly all of the Olympics' alpine glaciers most likely disappeared during the Hypsithermal period of global warming, the large glaciers on Mount Olympus may have lingered. Scientists were able to determine the approximate age of moraines left by the Blue

Mount Anderson and the recently deglaciated basin of the Anderson Glacier. Photographs taken early in the 1900s show the entire foreground covered with ice. Such evidence of the "Little Ice Age" can be found throughout the high mountains.

Janis Burger

Glacier on Mount Olympus by dendrochronology — counting the rings on trees growing on them. Early Neoglacial advances of the Blue date to around 1250 and 1650. Blue Glacier reached its maximum Neoglacial advance into the upper Hoh Valley in the early 19th century. By 1815, it had already begun its retreat up the valley, but for the rest of the century it would continue to cascade down a spectacular icefall below its present terminus.

The booming of the great river of ice down this cliff face, still visible in photographs taken in the early 1900s, may have given rise to a Hoh Indian version of the coastal legend of Thunderbird. It was believed that Thunderbird got angry when hunters approached his "house" at the foot of Mount Olympus. He thundered his complaint, sending ice cascading out his door and crashing down the mountainside. Not surprisingly, the legend goes, "no one would sleep near that place overnight." The legend was "confirmed" by observers visiting the area in the late 19th century, who reported low rumbling interspersed with booming crashes. The glacier began to retreat from the cliff face during the 1920s, and by the late 1930s the cliff was ice-free, as it remains today.

Studies of Blue Glacier began in earnest with the creation of Olympic National Park. Beginning in 1938, park naturalists measured and photographed the terminus of Blue Glacier every year. Other intensive studies launched during the International Geophysical Year of 1957–58 have continued to the present. The result — the longest record of continuous data for any glacier in North America — makes the Blue an excellent laboratory for studying relationships between weather patterns and the growth and shrinkage of glaciers.

Blue Glacier continued to recede from the late 1930s through 1953. But as a result of cooler weather and a heavier snowfall that began in the late 1940s, more than three decades of retreat, measuring more than 800 feet, ended when the glacier stabilized in 1954. The following year it took a half-hearted lunge forward, but by 1960 Blue Glacier had receded farther up the valley than it had since the early 19th century. With cooler weather and heavier snows in the late 1960s and early 1970s, Blue Glacier began to advance again, regaining more than 500 feet in length by 1980. The advance was brought to a halt by drought conditions that prevailed from 1977 to 1981. The glacier remained relatively stable for a few years; then in 1986, the Blue began a steady retreat. It has been receding at an average of more than 50 feet per year since then.

Current research on Blue Glacier seeks to correlate "mass wasting" (the amount a glacier loses by melting relative to its annual gain from precipitation) with local weather conditions. Since glaciers filter out short-term climatic fluctuations, only long-term data are preserved. Early results show a warming trend in winter air temperatures around the elevation of the glacier's terminus (more than 5 degrees Fahrenheit since 1948), while summer temperatures at upper levels of the glacier remain the same. This may indicate changes in the circulation pattern of the jet stream in winter — a condition that could prove to be a factor in current global warming.

No one knows where Blue Glacier — or the climatic changes that drive it — will go from here. Though evidence is mounting for increased worldwide warming, a brief look to the past shows that anything is possible. More accurately, anything is probable. In a region where the only climatic constant is change, it will be the genetic wisdom inherent in the plants and animals that have survived here, combined with the geographic diversity of the land itself, that will adapt and prevail. All else being equal, the ecosystem knows best what it's about.

As for us? If we humans are to survive the kinds of dramatic changes in world climate some scientists predict, it will be as we always have: by paying attention, reading the winds, and not being afraid to change our ways. Like most natural communities, human societies seem to do best when they reflect the diversity, adaptability, and tolerance of the elegant natural systems that surround them.

3
The High-Country Year

The air is cool, and patches of snow linger in the draws as I climb the trail through a mountain forest of western hemlock and silver fir. It's late spring back in the lowlands; the delicate trilliums and red currant blossoms have already begun to fade. But here, queen's cup has just opened its small white blossoms, and beadruby lifts slender clusters of cream-colored flowerets over leaves still damp with morning rain. It's this first freshness of spring that draws me to the high forest long before the mountain meadows open into the lush wildflower gardens of summer.

In the years I worked in these mountains, planting trees and clearing trails, I'd follow the advance of spring as the tattered edge of winter snow retreated up the mountain slopes. We'd start work in February most years, when the first blooms of Indian plum brightened the lowland forest. Late May would find us working the upper edges of the mountain forest. Above us, the meadows of the subalpine zone, where trees give way to open parkland and rocky cliff, were still deep in snow. Beyond that, the high open ridges of the alpine zone, well past the limit of trees, were raked by icy winds. Some years a late snow would sweep down and send

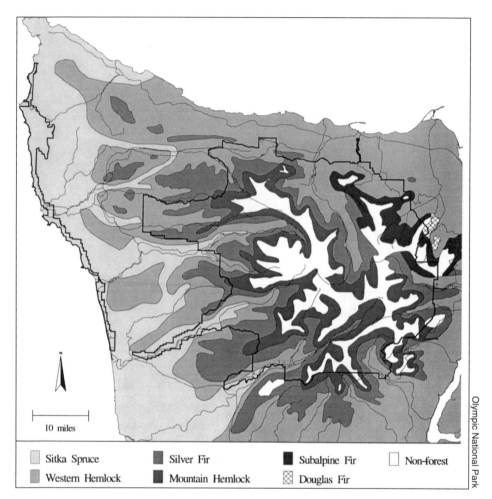

Sitka Spruce   Silver Fir   Subalpine Fir   Non-forest

Western Hemlock   Mountain Hemlock   Douglas Fir

Olympic National Park

The forest zones of the Olympic Peninsula.

us trundling back to the lowlands with our hats pulled over our ears.

No longer out on a daily basis, now I trace the season's progress by the bloom of wildflowers along familiar trails. Trailside plants along with subtle changes in forest composition also let me know how high I've wandered. Usually, at this time of year, it's the reappearance of the snow-pack that flashes the signal that I've entered the subalpine forest. But if it's later in the season, or the winter has been mild, I might wander past the first subtle shift into upper forest without seeing it. Then, somewhere around the 3,500-foot level, I'll look up and notice that the forest is grad-ing almost imperceptibly into the smaller mixed stands of mountain hemlock or subalpine fir that mark the lower reaches of the the subalpine zone, the portal of the Olympic high country.

High subalpine forests, sunlit wildflower meadows, and flashing snowmelt streams define the Olympic high country: snowfields and rock cliffs edged with gnarled scraps of forest, or wide rolling meadowlands broken by scattered copses of subalpine firs and stilled with alpine lakes. Here, as everywhere in the range, the beauty and richness of these mountains springs from the marriage of a varied and broken topography with the perennial gift of coastal weather.

Because of the heavy snowfall that buries the high country each winter, subalpine and alpine zones occur at lower elevations in the Olympics than in continental ranges such as the Rockies or the Sierras. These lower elevations and abundant moisture translate into less severe conditions for Olympic alpine communities: temperatures are moderate, ice storms rare, and drying summer winds less common. Even subtler factors such as soil temperatures and the degree of ultraviolet radiation are moderated by elevation and climate. All of this, combined with ample summer moisture from melting snow, has bequeathed to the Olympics' high forests and meadowlands an extraordinary lushness. And just as precipitation patterns have created diversity among the lower forests of the Olympics — from western rain forests to dry, rainshadow forests of Douglas-fir and lodgepole pine — prevailing weather patterns shape the nature of the Olympic high country. This becomes apparent to anyone who explores the high forests of Olympic National Park.

The Subalpine Forest

From Hurricane Ridge, the Bailey Range rises up out of the deep canyon of the Elwha River in a long southeasterly sweep of snowy summits. The range encloses the headwaters of the peninsula's major west-flowing rivers and, along with the high Olympus massif to the west, intercepts the brunt of weather systems moving in from the Pacific. Because of prevailing wind patterns, the mountains northeast of the Baileys receive much less rain and snow than the western and southern Olympics. At higher elevations throughout the range, where most of this precipitation falls as snow, the composition of subalpine forests and meadow communities and the kinds of wildlife they support are keyed to the amount of winter snow and how soon it melts in spring and summer. Other factors contribute, but the amount of moisture available during the short growing season shapes the alpine ecosystem.

West of the Bailey Range, the upper forests are dominated by mountain hemlock mixed with silver fir at its lower reaches. Alaska cedar is often found here, along with understory shrubs of oval-leaf huckleberry and white rhododendron. In spring, avalanche lilies pop up through receding snowbanks, and white-vein pyrolas and sidebells nod over damp mossy floors. Mountain hemlock *(Tsuga mertensiana)* is well adapted to wet snowy mountains and is commonly found in coastal ranges where winter snows exceed 10 feet. It is easily recognized by its full crown of dark green needles (uneven in length), stout limbs, and brown furrowed bark. Near timberline, mountain hemlock's gnarled, shaggy shape stands in bold profile against a snowfield or glacier.

In the drier eastern Olympics, the mountain hemlock–silver fir forests are replaced by more open forests of spirelike subalpine firs, with Douglas-fir and lodgepole pine occupying the driest south-facing ridges. Because of the lighter snowpack and generally more continental conditions on the east side (cooler winters and warmer summers), subalpine fir forests develop at higher elevations here — around 4,200 feet rather than the 3,500 feet for mountain hemlock — and they extend higher, to nearly 6,000 feet in elevation. In the central Olympics, habitats are mixed; subalpine firs favor south slopes while silver fir–mountain hemlock forests occur on the cooler north sides. Winter snow — and resulting summer moisture — affects tree growth differently east and west. Along Deer Ridge in the northeast Olympics, treeline is higher on north slopes where lingering snow nurtures summer growth, while along High Divide in the central Olympics, treeline is higher on south slopes because heavy snows inhibit tree growth on north sides.

Not surprisingly, the elevation of treeline is less than constant throughout the range; in fact it's hardly a line at all. Rather it's a mosaic of meadows and clumps and ribbons of trees that weave along peaks and ridges throughout the subalpine forest zone, either as small glades within groves of trees or as large openings. Some meadows result from lightning-caused fires and are maintained by heavy snows that kill tree seedlings. Heavy snows can also peel back forests along steep avalanche tracks.

Janis Burger

Subalpine fir copses alternate with meadows in the eastern Olympics. Copses result from layering, as lower limbs take root and grow new stems. In the western Olympics, mountain hemlock is the principal subalpine zone tree.

Thick tangles of Sitka alder and shrubs thrive in these areas between ascending tongues of subalpine trees. Together, the alpine and subalpine zones are a broad belt covering nearly a third of Olympic National Park. To a hiker, they are a montage of fluidly shifting environments that seem to change with each turn of the trail.

The emblematic tree of the high Olympics — as well as the high forests of most of the mountain West — is the subalpine fir *(Abies lasiocarpa)*. To breathe the rich balsamy scent of subalpine firs is to taste the essence of the high country. Extending from southeast Alaska and the Yukon as far south as New Mexico and Arizona, it is the most widespread true fir in North America, yet its elevational range is highly specialized. Unable to regenerate in the dense silver fir–western hemlock forests, subalpine fir pushes its upper ranges to the most windswept ridges. There, a few eke out a marginal existence as dwarfed and twisted *krummholz* shrubs ("crooked wood" in German), habitat for the hardy southern redback vole. Their more common spirelike form, however, is wonderfully well adapted for the cold snowy mountain environments in which they thrive.

The combination of snow, icy winds, a short growing season, and summer drought have given rise to some remarkable adaptations in subalpine firs. Notable at a distance across mountain meadows, the distinctive narrow symmetrical spires of these trees are well adapted for bearing up under heavy loads of snow. Their tough short limbs, thick with brushy blue-green needles, cluster densely around their trunks, shedding snow like a steep roof. Snow-laden limbs provide insulation as well as a protective curtain against blistering, wind-driven snow and ice. Subalpine firs are limbed to the ground, the lower limbs often forming skirts which, insulated beneath accumulations of snow, may take root in the moist duff-covered soil in spring, send up leaders, and form a cluster of new trees. This "layering" provides an additional means of regeneration not shared by mountain hemlocks or silver firs, and copses of close-growing subalpine firs are common at treeline in the Olympics. Surrounded by blue-leaf huckleberry, the copses provide shade and cover for blacktail deer as well as a number of bird species in summer. Over time, the parent trees die back, leaving ring-shaped clusters of firs scattered amid the mountain meadows.

Like Douglas-firs, subalpine firs are able to close off the stomates in their needles when the soil dries out in summer, shutting down photosynthesis and making themselves extremely tolerant of drought. But unlike Douglas-firs, they're also frost-tolerant, suiting them to alpine conditions. Though their low limbs and thin resinous bark make them susceptible to wildfire, they can regenerate successfully under a variety of conditions. During the warm Hypsithermal period between 6,000 and 10,000 years ago, subalpine firs were common throughout the Pacific Northwest; the high stands at treeline today may be remnants from that earlier time.

Traveling through subalpine fir stands in summer, it's common to hear the deep territorial booming of a male blue grouse from its drumming log. Grouse frequently nest beneath the branches of downed trees, nestling their clutches of brown-flecked buff-colored eggs in shallow scrapes. Later in the season, hikers can't help but notice the large cylindrical cones of subalpine firs: purple-hued and glistening with pitch, they seem to be cradled in nests of boughs. The understory shrubs in these forests include white rhododendrons on the north slopes and common junipers on the south slopes, with showy polemonium,

Scouler's harebell, and delicate starflowers brightening the forest floor.

Subalpine firs and mountain hemlocks are currently playing an important part in research on global climate change being conducted by the park service. Studies are underway to determine how vegetation might change in response to a changing climate. Scientists are monitoring the edges of alpine meadows where these trees are extending their ranges. Such ecotones (areas of transition between different environments) are natural laboratories; they can be seen as "tension zones" where plants test the limits of their habitats. Research has shown that different species in different microclimates extend their ranges during different climatic periods. For instance, during the first half of this century, milder winters and drier summers allowed mountain hemlocks to successfully colonize alpine meadows in the the western and southern Olympics. In this part of the mountains, where winter snowpack kills many tree seedlings, a period of less snow and warmer summers permitted mountain hemlock seedlings to become established in nearby meadows. Conditions in the northeast Olympics, however, proved to be quite different. Here, the major factor limiting the reproduction of subalpine firs is summer drought. Consequently, the firs didn't extend their range during the first half of this century, but beginning in the mid-1940s, increased winter precipitation and cooler, wetter summers allowed subalpine firs to advance into adjacent meadows. When conditions changed in the mid-1980s, the invasion ceased. Much like the study of alpine glaciers over time, the study of plants along ecotones like the upper edge of forest and meadow can help us learn much about the recent history of climatic change, and possibly assess future trends.

Mountain Meadow Communities

It's nearly summer by the time the high meadows of the Olympics begin to slough off their winter sheaths of snow and begin their short, intense growing season. Depending on the depth of the winter snowpack, snow melts back from different alpine environments at different times, creating a rich mosaic of blossoming meadowlands that unfolds along the mountain slopes throughout the short months of summer. It's a glorious time to visit the Olympic high country. It's a time of intense activity for

Janis Burger

One of eight plants endemic to the Olympic Mountains, Olympic rockmat blooms in midsummer and prefers clefts in steep rock faces. Seven more endemic plants grow only in the Olympics and on Vancouver Island to the north.

the small mammals that graze upon the emerging plants, as they forage for food in preparation for the winter and keep a wary eye out for predators — eagles, hawks, cougars, coyotes — that cruise the high meadows in search of them. Blacktail deer taste the morning mist from newly greening huckleberry shoots, and bees hum loudly among heather blossoms. Dark-eyed juncos arrive from the lower forests, working the edges of meadows for seeds and hawking for insects to feed their nestlings. Rare golden eagles cruise the cirque basins, alert for an unwary marmot, and gray jays and pine grosbeaks call among the subalpine firs. It's a brief interlude, barely five months from subalpine spring to fall — the first snows usually dust the high meadows by October, the first permanent snows fall by early November. But within this short window an unimaginable loveliness unfolds across the high country like an embroidered tapestry draped over a garden of stone.

More than 1,050 native plant species occur in the Olympic high country, including most of the Olympics' endemic and disjunct species. But these plants group themselves in different ways. Botanists working in the high Olympics have noted that certain plants tend to grow together as various microsites — open slopes, basins, swales, and such — emerge from winter snow. While keyed to the time of emergence from snow and availability of summer moisture, these communities are also shaped by elevation, aspect and degree of slope, soil depth, exposure to wind, and frequency of disturbance. Like their lowland counterparts, alpine and subalpine ecosystems are fluid and dynamic, making delineation of plant communities an inexact science at best. But by looking at plants in the associations where they frequently occur, we can get a fuller sense of the diversity and complexity of the Olympics' alpine areas.

After a mild winter, windswept alpine ridgetops and steep south-facing slopes around Hurricane Ridge and Blue Mountain begin to melt out by late April. The only real alpine tundra in the Olympics occurs in this northeastern corner of the range, most notably on Elk Mountain east of Hurricane Ridge. Except for these few isolated spots along ridges and mountaintops, the alpine zone is limited by the Olympics' relatively low elevation, the steepness of the high peaks, and the presence of permanent ice and snow. Lacking the high rolling ridges common to alpine areas of the Rockies, alpine plants in the Olympics are confined to narrow meadows along ridgetops and clumps of cushion plants on stable rocky slopes. On less stable scree slopes facing south or west, these plants tend to grow in long down-sloping ribbons or stripes.

Their early emergence from the snow exposes the plants that grow in these areas to extensive spring freezing and thawing, as well as severe summer drought. As a result, alpine plants have evolved unique strategies for coping with extreme conditions. Nearly all the flowering plants of the alpine and subalpine areas are perennials. Rather than growing new plant structures to reproduce by seed each year, they merely add to their plant tissue and survive quite handily, not flowering at all if conditions are unfavorable. Plants like the lovely spreading phlox that graces many of the Olympics' rocky slopes form cushions, as do delicate fronds of silky phacelia and the endemic Olympic Mountain synthyris. The dense low cushion shape minimizes exposure to the elements and insulates the plant against winds and extreme temperatures. Cushion plants catch and hold blowing dirt and debris which absorbs moisture and stabilizes their growing sites, as well as the seeds of other plants such as western wallflower and alpine timothy, giving them a place to grow.

Some plants that are only found in this community, such as smooth douglasia, grow in low mats. Less compact than cushion plants, mats commonly root by layering, which anchors them firmly to the loose gravelly slopes. Thread-leaf sandwort, mountain lover, and small-flowered blue-eyed Mary are also found in these early-emerging communities; so is one of the few nesting birds of the alpine, the horned lark. Horned larks have been known to build their nests beneath the cover of overhanging cushion plants, and their newly hatched young have been seen some years as early as June.

By mid-June, visitors to Hurricane Ridge will notice prairielike bunch grass meadows melting out along south- and west-facing slopes. Tufted thatches of Idaho fescue are soon followed by colorful blossoms of thread-leaf sandwort and bluebells-of-Scotland. Blue spires of broad-leaved lupine crowd the edges of subalpine forests here, and scalloped onion and spreading phlox colonize the lightly eroded areas. The Olympic snow mole, an Olympic endemic, can sometimes be seen here, and its diggings are evident among the grasses. Many of the plant species in this bunch grass community are also common to the sagebrush steppe of eastern Washington. Botanists suspect that this community, like the subalpine firs which border it, was much more widespread during the warm Hypsithermal period.

By late June in the northeast Olympics, even the shallow gullies and swales have become free of snow, but snowmelt streams and moist ground persist. The meadow communities that have developed here include conspicuous stands of American sandwort and cow parsnip, which partially shade understory plants such as waterleaf and pioneer violet, as well as ground creepers like American vetch. These robust multi-layered communities are also home to the only known alpine colonies of aplodontia or "mountain beavers" in the Pacific Northwest, and the meadows are frequently riddled with their burrows.

A similar but wetter meadow community develops on steep east- to northeast-facing slopes that melt out later in July. Dominated by Sitka valerian, which lifts a tall stem to a cluster of pink-white flowers with long lashlike stamens, showy sedge with its fuzzy flower spikes, and crowned with the subtle elegance of chocolate lilies, this may be the most elaborate of the Olympics' subalpine communities.

Along with these meadow communities, observant hikers can find a wealth of small wildflower gardens in microsites throughout the mountains. Small seeps provide habitat for the insectivorous common butter-wort. Parsley-fern clings to rock outcrops. Alpine willow-herbs and bright yellow monkey-flowers colonize wet streamsides while extravagant elephant's head and delicate bog orchid breathe life into pocket wetlands. Part of the joy of discovery while roaming the high meadows is recognizing old friends tucked away in unexpected neighborhoods.

As summer progresses, the high meadows and basins of the central Olympics melt out from the snowpack and a different, wetter series of meadow communities greets backcountry hikers. Among them, heather meadows are probably the best known — or at least the most easily recognized — subalpine plant community in the Pacific Northwest. The juniperlike foliage and small bell-shaped red and white blossoms of heather grace the high country of Mount Rainier, the North Cascades, and the Coast Mountains of British Columbia. In the lower-elevation subalpine areas of the Olympics, heather meadows are found mostly in the glacier-carved basins of north-facing cirques where conditions approximate those of higher ranges. Heather meadows are not expansive in the Olympics, but they occur throughout the range — from Grand Valley and Appleton Pass in the northern Olympics to Hart and Marmot lakes and upper Lena Lake to the south. The lush meadows of Seven Lakes basin below High Divide are among the richest heather meadows in the mountains. Here the sweet greens of avalanche lilies and blue-leaf huckleberry — tastiest of all huckleberries — abound, making them as popular with deer and black bears as with people. But don't search here for the inappropriately named heather vole; they're common in most other alpine habitats, but they don't much care for heather.

Given their dramatic settings in glacier-carved cirques, it's not surprising that the worst damage done to heather meadows comes from backcountry campers. The woody stems of heather hold up well under the weight of heavy snows, but continual trampling from lug-soled boots takes a toll. Olympic National Park managers are trying to restore damaged meadows by closing some sites to overnight use and raising subalpine plants from seed to revegetate damaged areas. The effort has been successful to some extent, but continued education about the nature and fragility of alpine communities and cooperation among increasing numbers of park visitors is necessary to protect existing and newly restored heather meadows.

Snowbanks linger longest in low swales and sinks, often within cool north-facing heather meadows. Many of these sites do not melt out until August, some not until September. The black alpine sedge meadows that develop here are usually small, and the plants low-growing. Sedges, pussytoes, thick fields of avalanche lilies, and the twin pink-lavender blossoms of alpine willow-herb are common here. Dusky shrews are known to favor sedge meadows, where they skitter about secretively in search of insects. The black alpine sedge community also has a wide distribution in North America, suggesting these plants were formerly more common. But in these post-glacial times, they persist only in the coolest sites on high mountains.

The most common type of meadow encountered by hikers throughout the Olympics is characterized by lush fields of tufted showy sedge. The singular white tufts of American bistort, bright yellow fan-leafed cinquefoil, and beds of lavender broad-leaved lupine flourish in the deep moist soils of showy sedge meadows. These meadows hug the north slopes and level ridgetops in the dry northeast Olympics, but elsewhere carpet west- and south-facing slopes. At Honeymoon Meadows on the West Fork Dosewallips, showy sedge grows waist high along the trail. In sites like this there's not much diversity; showy sedge outcompetes nearly everything else. Still, this is the most productive community in the subalpine, producing as much as six times more biomass than the bunch grass communities in the rainshadow. Showy sedge meadows are central to the ecology of the high country for another reason as well. Along with the bunch grass, valerian, and black alpine sedge communities, they form the heart of habitats preferred by the most gregarious and conspicuous of Olympic's alpine residents, the Olympic marmot.

The Olympic Marmot

Even after the snowiest of winters, when the mountain meadows are buried beneath May drifts of 10 feet or more, the one sure sign of spring is the skittery tracks of marmots newly emerged from their winter-long sleep. Their tracks are the first signs of life I see in the high country when I make early-season climbing trips, and the mere sight of them never fails to delight me. The sharp piercing whistle of a marmot's warn-

The Olympic marmot evolved into a separate species during the long isolation of the Pleistocene. Endemic subspecies in the Olympics include the Olympic chipmunk, the Olympic snow mole, the Mazama pocket gopher, and the Olympic short-tailed weasel.

Janis Burger

ing call seems to wake the whole subalpine community to life, and their loping gallop across the spring snow lets me know that the world still ticks true. Marmots are incredibly social mammals; their early tracks frequently lead to a spot directly over another burrow in the colony, then disappear into the snow. How they find the entrances beneath 10 feet of snow is a mystery, but they rarely miss.

The endemic Olympic marmot *(Marmota olympus)* is one of six North American marmot species — those large members of the ground squirrel clan — and the most social. They greet each other morning and night with nuzzling, nibbling, and play-wrestling. Extended families of about a dozen marmots share common colonies, feeding and resting together, exchanging sentry duties to guard against predators, and tolerating infants and yearlings much more than their cousins in the Rockies, the yellowbelly marmots. This behavior may be influenced by the long, seven- to eight-month hibernation period necessitated by the Olympics' heavy winter snows and the consequently short summer — during which marmots must mate, rear young, and pack on enough fat to sustain them through their winter hibernation. Maximizing cooperation and minimizing aggression seems a wise evolutionary imperative under such conditions.

A typical marmot colony is made up of an adult male, two adult females, yearlings, occasional two-year-olds and the summer's infants. Females do not mature until their third year and then breed only every other year, an adaptation that may reflect the short summer feeding period. Mating usually takes place in June, and the young emerge from their nursery burrows in late July. Ironically, mild winters cause the highest mortalities in marmot colonies, especially among the young. Winter snow must be deep enough to insulate their burrows during the long months of hibernation, and a healthy snowpack — and the abundant summer moisture it brings — is essential to their summer feed.

One of the many traits I admire in marmots is their fondness for wide front porches. Nearly every burrow complex is outfitted with one at its main entrance, and the sight of a marmot relaxing there in the morning sunlight brings to mind a cup of tea and the morning paper. Other burrows scattered across a typical five-acre colony afford quick escapes from predators. Most mornings are spent feeding in a group — the better for safety. Midday brings them back to their burrows for a long rest (another admirable habit) before evening feeding. As the summer unfolds, feeding activity increases.

Marmots have excellent taste in habitat. Nearly all marmot colonies in the Olympics are located in mosaic meadow complexes where three or more plant communities are intermingled. Marmots prefer newly emerging plants with tender immature flowers. These plants are higher in calories and proteins — and digestibility — than older plants. Since different plant communities emerge from the snow at different times, such mosaic habitats offer the highest productivity and the best chances for winter survival. Foraging marmots move systematically from bunch grass communities in June to showy sedge and valerian communities through July and August. By mid-August most years, marmots are wrestling less and concentrating their feeding activities in the late-emerging black alpine sedge communities, adding to their fat reserves every day. By October, when the meadow plants have died back, the marmots have safely sealed the entrances to their grass-lined hibernation burrows and snuggled in for another long nap. Barring a light snowpack or midwinter thaw, they'll wake to tunnel through deep snow in mid-May, popping up like scraps of cinnamon toast, ready for another summer's frolic in the high meadows.

More than any other alpine mammal, the highly specialized adaptations of Olympic marmots embody the coevolution of a native species with its environment. In the geographic and Ice Age isolation of the Olympic Mountains, the Olympic marmot evolved into a separate species with distinctive behavior and an ecology shaped by the nuances and seasonal variations of its subalpine environment. Similarly, the presence of marmots — burrowing and selectively feeding — has influenced what kind of plants grow in the meadows where they live; areas grazed by marmots support more species than ungrazed areas. These relationships, honed and deepened over millennia of adaptation and change, give us a glimpse of the wisdom and magic that lie at the heart of a living wilderness. Like the elk herds and their rain forest valleys, marmots are an inseparable part of their alpine world. By observing them, a trace of the elegant stitchwork that underlies the beauty of these mountains comes alive.

Natives and Non-natives

Though Olympic marmots are nearly ubiquitous in the high country, we know much less of the life ways of the other mountain-dwelling mammals endemic to the Olympics: the Olympic chipmunk, Olympic snow mole, the Mazama pocket gopher, and the Olympic short-tailed weasel. These animals have evolved into distinct subspecies in the isolation of these mountains, and the ecological webs and niches they occupy are a part of the fabric of this unique ecosystem. As important as these endemics in defining the Olympic ecosystem is the absence of a number of mammals common to the Cascades. These "missing dozen," as they became known, included grizzly bear, wolverine, lynx, pica, mountain sheep, mountain goat, and a number of smaller animals. In the past several decades, some of these animals, such as porcupines and coyotes, have colonized parts of the peninsula — the latter no doubt filling a niche left by the wolf, which was was hunted and trapped to extinction here by the 1920s. Such early blunders were a product of ignorance — the limited understanding of ecologic process of the day —combined with political pressure brought to bear by special interests, in this case local ranchers. But mistakes weren't limited to extirpating species from the ecosystem. Long-lasting problems and resulting damage to alpine and subalpine environments have resulted from a species introduction.

In the 1920s, before the creation of Olympic National Park, 12 mountain goats from British Columbia and Alaska were introduced into the Olympic Mountains. The goats flourished in the moderate coastal climate of the Olympics, and by early 1983, their numbers had increased to an estimated 1,200 animals. As early as the 1960s, goat damage in the form of extensive cropping of subalpine meadows and large barren wallows where goats dusted themselves became obvious. By the 1970s the goats occupied much of the alpine country of the park as well as adjoining Olympic National Forest. Most worrisome was that their range and forage requirements overlapped the ranges of several rare or endemic plant species, such as Piper's bellflower, Olympic Mountain aster, and Olympic Mountain groundsel. Because goats tend to overgraze their favorite plants, other plant species that favor disturbed areas were moving in, changing the composition of the park's alpine and subalpine plant communities. Since these communities evolved in the absence of large rocky-outcrop herbivores, including mountain goats, the goats clearly posed a threat to the ecological integrity of the park.

In the 1980s, the park service began an experimental management program that involved tagging, capture and removal, sterilization, and killing some animals for research purposes. Sterilization proved ineffective, and by 1990, live-capture became inefficient as the numbers of goats declined. Yet even one goat returning to a wallow seriously retards recovery of the habitat. In 1995, the park service released a draft plan to remove all the remaining non-native goats by the only effective means available, aerial shooting. The decision struck some as drastic and unwarranted, but the park is mandated to protect native ecosystems and manage exotic species.

Still, goats continue to live in the park as well as on adjoining forest service lands in the dry eastern Olympics, home of some of the highest concentrations of rare plants on the peninsula. Even if all goats are removed from the park, damage to fragile plant communities will persist on the forest service lands, and goats may move back into the park at will. If the ecological integrity of these communities is to be preserved, the bitter pill is that non-native goats must be removed from all jurisdictions in the Olympics, and alpine areas allowed to heal. Mountain goats are not a threatened species in Washington state, but the alpine and subalpine plant communities of the Olympics are irreplaceable.

Autumn in the High Country

The lid on my water bottle was frozen shut, and my cooking stove was rimed with frost. But morning sun already reached the far edge of the meadow, stirring mist like steam from slowly warming broth. Yesterday, the edges of a high lake near camp still held ice in mid-afternoon, and last night's rain whitened the ridgetop above me. This was my last visit to this high basin before winter locked the country in white, and I drew out my stay as long as I could. I didn't want to leave without a last glimpse of the elk that had been browsing this basin throughout the late summer and fall. I had seen fresh sign everywhere, but for the past few days the elk had eluded me, and I was afraid they had already returned to the low-lands. One more range through some high basins and I would head down too, I told myself, reluctantly as always. It's a crime to leave a camp like this in good weather.

In late summer and fall, the high forests and meadows fairly ring with the bugling of Roosevelt elk. As soon as the high country is free of snow, elk migrate up from surrounding valley forests to partake of the summer growth. By fall the bulls have herded their "harems" together, and their bugling echoes across cirques and basins. Black bears have also come to the high meadows to fatten themselves on the huckleberry crop, and Cooper's and sharp-shinned hawks have left their nesting areas in the deep forest to join other hawks, larks, and owls feeding on the late-season flush of band-winged grasshoppers. In these creatures the hands of the forest reach up to gather the bounty of the high sunlit slopes, before drawing back into winter. In them, the singular wholeness of the ecosystem shines through.

Fall is my favorite time in the Olympic high country. There's a crisp cleanness to the air, and fresh winds rustle the seed pods of summer wildflowers. It's a time when the meadows exchange the exuberant hues of summer blooms for the burnt sunset colors of fall. As peninsula botanist Ed Tisch has written, "At upper levels . . . where open forest gives way to meadow, the arrival of autumn is not simply the end of summer, but a brilliant culmination where color streams from dying leaves and straggling flowers."

Now that brilliant culmination was upon me. The leaves of white rhododendrons along the upper edges of treeline had turned golden with the first frosts, and mountain ash leaves kindled a bright fiery orange. Closer to the ground, the leaves of thin- and blue-leaf huckleberries ran from burgundy to a stunning bluish rose, a color matched only by the twilight sky. The sheer intensity of the summer season burned on in the alpine fall like the embers of a dying fire.

On my last day in the high country, I followed an elk trail that traversed a series of high basins. Toward noon I stopped at a windy pass and finished my lunch as the inevitable clouds began to build against the summits above me. The open basin below was empty of wildlife; not even the whistle of a marmot broke the windy silence. Just as I made ready to go, I noticed some movement in the timber far below. And there, across a low finger of meadow, came a large bull leading a band of elk. The bull waited while the elk followed across the narrow bottom of a cirque, then moved singly and by twos and threes up a steep talus slope and back into a fringe of trees. Through my field glasses I tried to count them, but ultimately gave up and watched as they disappeared, a river of buff rump patches swallowed up by the arms of the forest. Then, like a watchman closing a door, the bull turned and stepped slowly into the trees.

4
The Rain Forest

Spring comes early to the rain forest valleys of the western Olympics. That may be part of their magic. Tucked beneath a blanket of winter clouds and rain, lush beds of mosses and liverworts — which cover nearly every horizontal surface here — awake and begin their growing season. But theirs is a subtle bloom, seldom noticed. More vivid are the small blossoms of spring-beauty and pioneer violet that glint from damp forest floors some years as early as March. By April, curled white petals of trillium float leaflike and immaculate above the moss and oxalis, and soon, the cream-white blossoms of bunchberry dogwood fringe the buttress roots of huge rain forest trees.

Even amid the full burst of spring, it's the trees that capture visitors' attention. Giant Sitka spruce, western hemlock, western redcedar, and Douglas-firs lift grandly into a cathedral canopy of green and broken light. There's something in the stature and grace of these old trees that never fails to slow my pace and alter my perspective. To enter an ancient forest is to step into an ongoing story as old as life itself. Breathing the sweet oxygen-rich air, something inside me knows I'm part of that story, and I recognize the forest as home.

The early maple leaves lend a golden hue to this moss green world, and scarves of clubmoss and aerial fronds of licorice fern sway gently in the valley breeze. The rain forest is awash with life. Even the moss-carpeted logs that litter the forest floor support gardens of blooming wildflowers. Dense thickets of forest seedlings crowd the broad shoulders of other logs, suspended above the thick growth of the forest floor like stranded sailors clinging to a life raft.

The temperate rain forest is the most spectacular of the old-growth forests that mantle the lowland valleys of Olympic National Park. It represents a coastal forest type that stretches 2,000 miles from southern Oregon to southeast Alaska. The Sitka spruce–western hemlock forests that dominate these low-lying coastal areas, like the redwoods that grow to the south, are bound by a narrow coastal fog belt that rarely extends more than a dozen miles inland. That this forest type reaches its maximum potential in the inland valleys of the western Olympics is the fortunate result of a number of factors. Over time, they have given rise to a magnificent forest community, one that boasts the greatest biological productivity in the world.

The stage for the Olympic rain forest was set during the Pleistocene, when alpine glaciers repeatedly bore down west-flowing river courses, carving deep, flat, U-shaped troughs. As glaciers retreated, valley floors were filled with glacial rubble and debris. The deep, long valleys that resulted — the Hoh, Queets, Quinault, and to a lesser extent the Bogachiel — were aligned with the prevailing southwest winter storm track. In consort with the abrupt rise of their surrounding mountains, these valleys funnel prodigious amounts of precipitation into their upper reaches. While Kalaloch on the coast receives 90 inches of precipitation each year, the Hoh Ranger Station, less than 30 miles inland, molders beneath 140 inches. A dozen or so miles farther into the mountains, Mount Olympus tops the charts with more than 200 inches. As this precipitation pattern held during the climatic fluctuations that followed the Pleistocene, alpine glaciers have continued to pulse forward and melt back. Drift from these recent glacial advances has left a series of terraces in the valleys. Outwash from meltwaters as the glaciers retreated carved through them, resulting in a series of gentle stepped terraces along valley floors. The age of these terraces in turn influences the types of vegetation they support. The mix, from young disturbance species along gravel bars and bottom lands to the ancient stands of upper terraces, offers a diversity of habitats for the rain forest's numerous wildlife species.

During the brief dry months of summer, blankets of coastal fog funnel up rain forest valleys and are strained through the myriad needles of tall conifers to fall onto forest floors as "tree drip." This booster shot of summer moisture, which can add as much as 30 inches of precipitation to the forest annually, moderates summer drought and reduces the chance of serious wildfire. Steep, glacier-carved valley walls also provide protection from the windstorms that ravage coastal forests elsewhere. This mix of protection from natural disturbance and an optimal climate has allowed the Olympic rain forest to achieve great age and complexity. And it's produced some exceptionally large trees.

Trees of the Rain Forest

Sitka spruce *(Picea sitchensis)*, the tree most emblematic of the rain forest, has sharp bluish green needles that grow brushlike from the twig, and scaly purplish brown bark. Mature spruces rise from their fluted bases in straight untapering trunks to heights of up to 300 feet. Spruce trees five to eight feet in diameter are not uncommon in the rain forest. A large spruce near the Hoh Visitor Center is 13 feet in diameter, and one rain forest giant growing near Lake Quinault was measured at nearly 19 feet. Other tree species also reach huge dimensions here. The largest Douglas-firs in the world grow in the Queets Valley, and the four largest known western hemlocks are found in the western valleys of the park. The current record redcedar, more than 20 feet in diameter, grows on the north shore of Quinault Lake. Of course these records change as new trees are found and former champions lose their tops or fall. But given the extent of past logging in the peninsula's coastal forest, most of it along the productive, low-elevation coastal plain, it's doubtful that the true record trees were ever measured.

Sitka spruces, like the western hemlocks that share dominance in the rain forest, are shade tolerant. Both are able to reproduce beneath the shade of the old-growth canopy. Because of this, they seed in readily beneath early-succession trees like red alder or Douglas-fir, which depend on freshly disturbed ground and sunlight. Eventually, spruces and hemlocks come to dominate older rain forest stands. Lacking major disturbances such as windstorms or fires, the spruce-hemlock community can perpetuate itself indefinitely. But disturbances do occur; trees are toppled by windthrow or succumb to insects or lightning fires. That's when Douglas-firs seed in to newly exposed soils. Less common than spruce or hemlock in the rain forest, Douglas-fir is unable to reproduce in forest shade. The towering firs that grow along scarps and terraces stand as witness to fires, windstorms, or river cuts that disturbed the forest centuries ago. Sitka spruce can also survive on freshly disturbed soils, but in deep forest shade, western hemlock has the edge over spruce. Eventually, hemlock comes to dominate most lowland forests elsewhere

Sitka spruce is a fog-loving coastal tree that grows along a 2,000-mile belt from Oregon to Alaska. It reaches its greatest development in the temperate rain forest valleys of the Olympics; spruces here have attained diameters of 18 feet and heights approaching 300 feet.

Janis Burger

in the Northwest. That a coastal species like spruce continues to hold its own in these inland valleys no doubt is due largely to generous amounts of rain and the coastal fog, but another factor in spruce's success may be elk. The resident herds of Roosevelt elk that browse these valleys shun prickly spruce seedlings, but they go for hemlocks in a big way.

Other trees are also woven into the fabric of the rain forest: red alder, northern black cottonwood, bigleaf maple, and western redcedar are most noticeable. Where and how thickly they grow in the forest can tell us much about how the rain forest community develops over time. In the near absence of major windstorms and fires, the prime disturbance factor in these forests may well be the rivers. As they braid and meander across their floodways, rain forest rivers carve at forested banks, shore up new gravel bars, and flood bottom lands, leaving behind layers of fine sandy soil. These activities intensified over the past thousand years with the glacial advances of the Little Ice Age. The resulting valley terraces and the plant communities they support offer a glimpse into the stages of rain forest succession — the procession of plant species through time that leads toward stability or "climax."

River Terrace Communities

In the Hoh Valley, a sequence of communities on progressively aged river terraces corresponds with periods of recent glacial advance. To wander across these terraces, from the river bank to valley walls, is to pass through the successional stages of the temperate rain forest and see how this ancient community develops over centuries, from simple pioneering communities to the complex structure of an ancient forest.

Red alder and willow thickets begin to crowd young gravel bars along the river within a few years after they form. As the river migrates back across its floodway, mature alders come to dominate these areas; grasses carpet the understory and Sitka spruces begin seeding into the young soils. These alder bottoms are favorite camping spots along the river; they also serve as important feeding areas for valley elk, particularly during spring and summer.

The first terraces, just above the alder flats on the bottom lands, are slightly older, having been deposited by the river about 400 years ago. Here, the spruce seedlings of the alder flats have grown into mature stands, while the short-lived alders have dropped out. Western hemlocks also appear on these terraces, mostly as smaller trees and seedlings on fallen logs. Bigleaf maple and black cottonwood trees are abundant, and a few older Douglas-firs linger from earlier disturbances. The grasses that dominated the alder flats are now joined by shamrock-like oxalis, enchanter's nightshade, sword fern, and mosses, and low spreading limbs of vine maples seek sunlight in the dense forest shade.

On the second terraces, older trees are dying, allowing the canopy to open. Sitka spruce and hemlock are codominant here, forming a climax community. The spruces here are 500 to 600 years old. Vine maples grow in thick tangles, and salmonberry and huckleberry shrubs are common. Sun-loving grasses have all but disappeared now, and ferns, mosses, and beds of oxalis cover much of the ground. Mossy groves of big-leaf maples occupy the disturbed sites of old landslides and the alluvial fans of side streams, where they're joined by the graceful fluted columns of western redcedars.

The third and oldest terraces are closest to the valley walls. Laid down by the alpine glaciers of the Pleistocene more than 15,000 years ago, they support the deepest soils and the oldest valley forest communities. Most of the trees are ancient hemlocks, and a low growth of oxalis, foamflower, bracken, and moss blankets the ground. Beyond them, valley walls host an entirely different forest type, where rain forest communities give way to montane forests of the upper slopes.

Nurse Logs and the Forest Floor

Among the signature characteristics of the temperate rain forest — large tree size, abundant mosses, lichens, and liverworts, and multilayered canopies — is an abundance of dead and downed logs. Many of these serve as "nurse logs," fallen trees that act as nurseries for forest seedlings. Seed germination on the rain forest floor is scant because of the thickness of the moss and herb layer. Seedlings either can't compete, or fallen seeds remain suspended above the soil. Nurse logs offer a head start for seedlings, providing nutrients, moisture, and in some instances, protection from browsing elk. Nearly all the trees in the rain forest had their beginnings on nurse logs, upturned root wads, or broken-off stumps.

Soon after a tree falls, it's attacked by bark beetles and wood borers. Entering first the soft tissue of the cambium layer and later boring into sapwood and heartwood, these insects, and the microbes and fungi that accompany them, are pioneers, initiating a complex community of consumers, predators, and scavengers which over decades will return the nutrients and organic materials of the fallen tree to the soil. Sitka spruce and western hemlock decompose relatively quickly, but Douglas-fir and western redcedar can last almost as long on the forest floor as they stood in life — 700 or 800 years for a fir, a thousand for a redcedar. Throughout the process of decomposition, downed logs provide habitat for a plethora of organisms from bacteria and fungi to spiders, mites, millipedes, and snails. Even vertebrates such as salamanders, shrews, and voles use downed logs for feeding, cover, and travel routes over the crowded forest floor. The complex interactions between plants, animals, and nutrients in decomposing logs is a critical part of the energy cycle of old-growth forest ecosystems.

While most of these functions are hidden from view, nurse logs provide a poignant glimpse of new life springing from old. The moss growing in the bark crevices of newly fallen trees creates a secure and moist environment for tree seeds. Hundreds if not thousands of seedlings may crowd along the length of a nurse log, but only a few will develop into mature trees. Their survival ultimately depends on their roots reaching the soil. Some roots gain advantage by finding their way into insect borings and tapping the moist, decomposing wood inside the nurse log. But these early starters face intense competition from other seedlings, as well as crowding by the deepening moss layer, sloughing bark, and toppling by snow and wind. Browsing elk and summer drought can also take a toll. Within a decade, only one in ten seedlings will survive; only one in thousands ever becomes an old-growth tree. Still, evidence of trees' nurse-log beginnings is apparent throughout the rain forest, from "stilted" trees which had their start atop stumps and root wads, to rows of mature trees or "colonnades," roots still clutching the hollow ghost of a vanished nurse log. There are few places where the eternal cycle of life drawing from death is more strikingly apparent.

The Roof of the Forest

While the cycle of life is easily seen in nurse logs and colonnades, much of the life of the rain forest is less apparent to visitors because it goes on above their heads. A profusion of mosses, lichens, liverworts, and ferns drape the limbs and trunks of rain forest trees. Virtually all canopy plants are epiphytes, plants that grow on other plants without parasitizing them. Drawing their nutrients from the atmosphere, rain, blowing dust, and fog drip passing through the canopy, these epiphyte communities serve a number of functions beneficial to the health and maintenance of the forest ecosystem.

Easily the most noticeable of these communities are the luxurious hanging mosses that drape the limbs of bigleaf maples. These aerial gardens, which form moss mats six to ten inches deep, are composed primarily of the clubmoss *Selaginella oregana*. Closer in structure to a fern than a moss, selaginella is joined by sphagnum and other mosses, as well as liverworts, lichens, and ferns. With a collective weight of a ton or more per tree, these moss mats have twice the biomass of the understory plants that grow on the forest floor and are four times heavier than the trees' own foliage. But the bigleaf maples and other rain forest trees do not merely serve as toeholds for these epiphyte communities. Scaling climbing ropes into the canopies of bigleaf maples, researcher Nalini Nadkarni discovered tree roots growing in the organic soil that had developed at the base of epiphyte mats. The large bigleaf maples, and vine maples and cottonwoods to a lesser extent, tap into their moss mats with roots identical to the roots they put down into forest soil. This strategy enables them to garner extra nutrients from these newly forming soils, and completes a symbiotic relationship between the trees and their epiphyte gardens.

Throughout the old-growth forests of Olympic, the canopy is home to a rich assortment of invertebrates, birds, and small mammals, as well as a complex community of microorganisms whose importance to the health of the forest ecosystem is only now beginning to be understood. More than 130 epiphytes have been identified in the Hoh Valley alone, the great majority of them lichens. Lichens are formed of a partnership between fungi, which provide structure, and algae, which photosynthesize sunlight. One group of canopy lichens provide the valuable function

Janis Burger

Epiphyte mats and licorice ferns. Such canopy plants do not harm their host trees. In fact, some rain forest trees may benefit from these epiphyte mats by sprouting roots into them and gaining extra nourishment.

of "fixing" essential nitrogen from the atmosphere and making it available to the forest ecosystem. These cyanolichens, as they are called, are capable of producing two to six pounds of nitrogen per acre each year — a considerable portion of the total nitrogen needs of the forest. Some cyanolichens, like the common *Lobaria oregana*, are blown down from the canopy in winter storms and provide a nutritious and easily digestible food for elk and deer. Another group, the Alectorioid lichens, are extremely sensitive to air pollution and serve as excellent indicators of air quality. Chemical analysis of some of these lichens in the Hoh Valley indicate that the Olympic rain forest has some of the cleanest air in the world.

Various groups of lichens occur at different stages of forest development. For instance, important cyanolichens and Alectorioid lichens do not appear until younger forests begin to age, and they do not dominate canopies until forests reach 400 years. Lichens and other epiphytes also arrange themselves in zones in the forest canopy. Fog drip, rain, and snowmelt filtering through the canopy as "throughfall" becomes chemically altered as it leaches through each successive layer. The resulting solution, a nutrient-rich stew loaded with nitrogen and phosphorus, is an important source of nutrients for forest plants including the trees themselves.

The old-growth forest canopy has been called the "last scientific frontier." Early research suggests that a wealth of new invertebrate species are likely to be discovered there. Studies conducted in the Quinault Valley found that arthropod communities — mites, springtails, millipedes, and spiders — are as numerous in the canopy as in the forest soil. A more extensive study conducted in a similar forest in Vancouver Island's Carmanah Valley may result in the discovery of dozens of arthropod species new to science. We're accustomed to hearing of this kind of diversity in tropical rain forests; that our own temperate forests could also contain this level of richness should be reason enough to preserve the last unprotected remnants of old-growth forests. If we fail, we may never know the extent of what we've lost.

The Rain Forest Elk

In fall, the rain forest valleys echo with the high whistling calls of elk. During the fall rutting season the great bulls stomp and pace back and forth through the forest clearings, warding off challengers and attempting to keep their "harems" intact. This is my favorite time to watch them, keeping downwind, in the cover of nearby trees. Their tan, muscular shoulders, polished antlers, and blond rump patches flash among the subdued greens and scarlet-edged golds of the autumn forest. The darker fur of their necks and legs blend with the shadows of forest trees. Where shafts of sunlight penetrate the canopy, their breath explodes in clouds of steam. Elk are known to have been a part of these forests for at least 3,000 years, and evidence of their continued presence, from scat and tracks in churned-up mud to browsed shrubs and ferns, is everywhere you look. Though they can be elusive in summer, when some of the herds migrate to high meadows and others favor the shade of dense growth, the frenzy of the autumn rutting season brings them front and center. For me, it's difficult to think of these valley forests without them.

The Roosevelt elk of the Olympics *(Cervus elaphus rooseveltī)* are the largest of North American elk and the second largest animal (after moose) in the deer family. Bulls can weigh up to 1,000 pounds, or about four times the size of the blacktail deer that share their range. Olympic National Park protects the largest population of Roosevelt elk remaining in their natural habitat.

Nonmigratory elk in the Hoh Valley, those that stay year-round, typically live in herds of 15 to 25, occupying home ranges of four to ten square miles. After a summer of feeding on the lush growth of the forest floor, elk are at their peak condition. Following the fall rutting season, which lasts four to six weeks in September and October, the bulls disperse and the elk return to their matriarchal social patterns. Herds commonly include older cows, their female offspring, and their calves. Upon reaching maturity, bulls generally drift away singly or in small groups.

The forest provides winter food in the form of hemlock boughs, alder buds, sword fern, and lichen, as well as cover and shelter from winter storms. During periods of heavy snows, elk may leave their favored valley bottoms and gather nearby on low south-facing slopes where warmer temperatures and less snow mean better forage. In February and March, herds begin to frequent alder bottoms where increasing sunlight ushers in the first spring growth of plants. Calves are born in May and June when food is abundant. Migratory herds, those usually occupying the upper portions of valleys, make their way into the high-country meadows in the weeks that follow. Herds typically stay together throughout the summer months, either in subalpine range or in valleys. By September the bulls have grown new antlers and once more begin to assemble sizable harems and sound their shrill and guttural calls across meadow and forest.

When I view these animals in their natural environment, it strikes me that the tremendous productivity of these forests has contributed to the Roosevelt elk's impressive size and stature. More intriguing is the way elk have shaped the character and structure of the rain forest. In several dif-

ferent studies, biologists attempted to quantify elks' effect on forest vegetation by constructing fenced "exclosures" that prevented elk and deer
from browsing small samples of valley forest. Two larger exclosures were
built in the South Fork Hoh Valley in 1980.

The contrast between vegetation inside and outside the fence is stunning. Outside, the open grassy "parklike" character of the forest floor is
broken only by low clusters of ferns and shrubs. Inside the fence is a jun-
gle. Salmonberry and thimbleberry bushes grow profusely, as do
huckleberry and waist-high sword fern. Many herbs that grow small and
sparse outside the exclosure loom large and vigorous inside, and the
grassy glades that roll to the fence line are nowhere to be found within it.
Young saplings such as vine maple and hemlock are much more numerous inside the exclosures than in the surrounding valley forest.

In 1990, park service biologists conducted a survey of 22 older exclosures along with the two built in 1980. Surprisingly, they found that
there were actually more species growing outside the exclosures than
inside, and the total cover of grasses and forbs was much greater. By limiting the growth of shrubs and ferns through browsing and trampling,
elk help create and maintain the groomed appearance of the forest floor.
This, in turn, results in an increase of the grasses and forbs that elk, as
well as other herbivores such as deer, mice, voles, and snowshoe hares,
prefer for food. This is particularly true on the floodways and younger
river terraces most frequented by elk.

But the story doesn't end there. Along with shaping the character of
the forest floor, browsing elk may also influence the structure of the forest itself. Observations of the South Fork Hoh exclosures suggest that
the abundance of Sitka spruce in the rain forest may be due in part to
browsing elk's taste for hemlock seedlings and avoidance of Sitka spruce.
This kind of mutuality in an old-growth forest system is striking. The
coevolution of the elk and the rain forest is yet another window into the
complexity of a pristine wilderness ecosystem. Even when no elk are to
be seen, their presence permeates the beauty of the rain forest.

It wasn't too long ago, however, that this story nearly took a sharp turn
for the worse. By the turn of the twentieth century, the Olympics' elk
herds had been decimated by overhunting. Elk were killed first for their
meat and hides; later, they were shot purely for their upper canine teeth,
prized as watch fobs by members of the Benevolent and Protective Order

of Elks. By 1905, the estimated elk population on the peninsula was reduced to less than 2,000 animals. Public outcry over the carnage led to President Theodore Roosevelt's creation of Mt. Olympus National Monument in 1909, which protected a core area of elk habitat in the central mountains. This, along with earlier hunting restrictions by the State of Washington, allowed elk numbers to rebound. By the mid-1930s, wildlife biologists were reporting "overbrowsing" of the peninsula's elk range. Extended elk hunting seasons were opened again in 1936.

The following year, 4,000 hunters descended on the western peninsula. More than 800 elk were killed, many left lying in the woods. Two hunters were also killed in the melee, and several more were injured. The excesses of the 1937 elk hunt proved a powerful argument for the creation of Olympic National Park in 1938. The park protected much more of the elk's rain forest habitat than the original monument, and it remained closed to hunting. By the 1950s, the Olympics' elk population had stabilized at its present level of about 5,000 animals. Some 3,000 thrive in the park's western rain forest valleys, a number well within

One of the most conspicuous inhabitants of the rain forest, Roosevelt elk may help shape the character of the forest as well. The open, parklike appearance of rain forest valleys is largely the result of browsing by elk herds.

Janis Burger

the carrying capacity of their environment. This is somewhat less true for elk that inhabit the Queets River corridor, a narrow band of protected river valley closely bounded on both sides by managed state and private timberlands. And it doesn't hold true at all for migratory east-side elk populations.

The small bands of elk that inhabit the steep, narrow valleys of the Gray Wolf–Dungeness, Dosewallips, Duckabush, and Skokomish rivers that drain north and east into the Strait of Juan de Fuca and Puget Sound spend only part of each year in Olympic National Park. The Dosewallips and Duckabush herds are known to migrate 15 to 20 miles, climbing nearly 5,000 feet in elevation to reach summer ranges in subalpine meadowlands of the park. In winter, however, these animals must descend the mountains and leave the park to forage on adjoining forest service, state, and privately owned timberlands and valley farms. Large tracts of this winter range have succumbed to logging and residential development over the past two decades, and hunting pressure has increased dramatically. As a result, east-side elk numbers have plummeted.

Currently, a moratorium on hunting, combined with a cooperative agreement between Indian tribes, the state, and private timber companies to manage timber cutting to preserve critical habitat, offers some hope. But whether this eleventh-hour truce will save the day for the park's east-side elk remains to be seen. Studies of elk that use logged lands as well as park lands along the Queets corridor on the west side have shown that elk generally avoid the dense, even-aged plantations that replace natural forests in the wake of clearcut logging. Rather, they prefer old-growth forests along with younger clearcuts that somewhat mimic young alder flats along valley floors. Much like the east-side elk, the long-term survival of the Queets corridor elk may depend on sensitive management practices on state and corporate timberlands outside Olympic National Park.

Cougar and Wolf

Elk and deer are major prey for that most secretive and solitary hunter of the Olympic forests, the cougar. Cougars are the "ghosts" of the forest, reclusive and silent. Unlike the more visible black bear and coyote, which are primarily scavengers of winter-killed elk and deer, cougars usually shun carcasses, preferring to take their meat on the hoof. Though little is known of the cougar population in the Olympics, evidence of their presence — tracks and the scented scratch mounds in the dirt that mark their territories — is plentiful on backcountry trails, and they are believed to be quite common. In more than two decades of wandering those trails, I've seen only one, but it was an unforgettable experience. A large, tawny, and singularly graceful animal with distinctive facial markings, it slipped only part way into the brush ahead of me, then stopped briefly to peer back before disappearing into the forest.

Cougars, or mountain lions (*Felis concolor*), are highly skilled and efficient hunters, perhaps the most adept large carnivore in North America. Unlike wolves, who hunt in packs, a cougar hunts alone, advancing on its prey stealthily, with slow, fluid, snakelike movements, before launching a carefully timed attack. It reaches its quarry in a bound or two, pouncing on the animal's back. A single gripping bite to the neck with the cougar's long incisors and powerful jaws can suffocate an elk or deer or sever its spinal cord. Though such acts of predation are rarely observed in the wild, a researcher tracking a radio-collared cougar in the Rockies arrived at a fresh kill site where the 150-pound cat had just taken down a 1,000-pound bull elk. The elk's neck was broken and its antlers driven nine inches into the ground.

The antlers and hooves of a large elk can inflict crippling or fatal injuries to cougars as well, however. There are numerous reported instances in which an elk was wary or a cat's timing was off, and the battle went the other way. Thus, where cougars prey on elk herds, they usually don't select quarry at random. Rather, they tend to go after more vulnerable or exposed animals, the tired or weak, sick, aged, or young. A cougar can take on average one large animal every 12 to 14 days, more, perhaps, when a mother is feeding kittens (a growing family can devour a deer overnight). Though such persistent hunting does affect populations of elk and deer, cougar predation is not usually a major factor determining herd size. The amount of habitat and quality and availability of winter range are far more important.

Young cougars spend most of their first two years with their mother. They're nursed for the first eight weeks, then taken to kill sites to eat small amounts of flesh. Gradually they learn how and what to hunt as they sharpen their skills in preparation for a solitary life as an adult. After 12 to 20 months, most young cougars disperse, often over long distances, and find their own territories, a process that may take years. During this transient period they are especially vulnerable. They may be killed by other cougars defending territories of their own, or as has been more common recently, they may wander into areas occupied by people.

Several encounters between humans and young adult cougars on the peninsula have ended badly for the cougars in recent years. As the human population increases and more people move into previously wild areas, cougars may be losing some of their instinctive fear of humans. In 1994, Olympic National Forest reported its first instance of a young cougar attacking a child. Fortunately, the child's father and other adults chased the cat off. The only documented instance of a cougar fatally injuring a person in this state occurred more than 70 years ago. An earlier report of a cougar killing a child near Sappho on the peninsula dates from 1897.

Still, incidents such as these have been increasing throughout the West, and park rangers are now advising visitors hiking with children to keep them close by. Hikers who come upon a carcass that might be a fresh kill are advised to give the area a wide berth. Those few who may suddenly meet a cougar should not flee, but stand their ground or back slowly away, giving it a chance to leave. As forest lands outside the park are converted to residential development and civilization continues to encroach into formerly wild habitats, such encounters between humans and cougars will likely continue. And as is inevitably the case in such trends, it will almost certainly be the wild animals that pay the price.

Historically, the Roosevelt elk and blacktail deer in the Olympics were also preyed upon by another efficient though less solitary predator, the wolf. Before settlement, gray wolves coexisted with elk herds in most valleys of the Olympics. Early reports described them as medium-sized wolves with reddish legs. They were most often reported hunting in small groups but were sometimes sighted alone. Unfortunately, the details of their social behavior and their relationship to the elk herds they hunted will never be known.

Not long after settlers arrived on the peninsula, they began an aggressive and systematic campaign to eliminate predators from the Olympics.

Wolves were hunted with hounds, trapped, and poisoned with strychnine-baited carcasses. Up to a million dollars a year was paid in bounties throughout the West during this time. In the Olympics, as elsewhere, the effort was chillingly effective. By 1917, when the U.S. Biological Survey sent Olaus Murie to the Olympics to eradicate wolves, he resigned having found not a wolf. Two years later, a single wolf was trapped near the upper Gray Wolf River. In 1920, a settler in the Elwha Valley trapped the last documented wolf in the Olympics. By the time the park was established nearly two decades later, there were no wolves left to save.

At one time, wolves inhabited most of the North American continent. Healthy populations still exist today in Alaska, Minnesota, and Canada. A population of *Canis lupis fuscus,* the subspecies that inhabited the Olympics, persists in coastal forests of British Columbia. In the U.S. the gray wolf is listed under the Endangered Species Act as endangered in all the lower 48 states except Minnesota, where it's listed as threatened. The act directs federal agencies to insure the preservation and recovery of any species threatened with extinction in "all or a significant portion of its range." As early as 1935, wildlife biologist Adolph Murie proposed reintroducing wolves to the Olympics. A 1975 feasibility study concluded that Olympic National Park could support a population of about 50 wolves. And in 1981, a National Park Service advisory committee nominated Olympic as one of the two best wolf reintroduction sites in the national park system. The other site, Yellowstone National Park, has recently embarked on a program of wolf reintroduction as part of a northern Rockies wolf recovery project. Wolf restoration there has been a remarkable success.

A 1991 headline in a local newspaper typifies local opposition to wolves. "Livestock losses, timber reductions possible if grays return." The accompanying story quotes area ranchers' concerns about stock losses due to wolf predation. Their fears may be exaggerated. In Minnesota, where some 2,000 wolves range near dairy farming country, less than one in 10,000 cows are taken by wolves, and in Montana's sheep country, dogs kill far more sheep than wolves do. A 1982 amendment to the Endangered Species Act allows ranchers to kill wolves that take livestock, and most biologists understand that active management of wolves must be part of any recovery program. In spite of this, the same fears and uncertainties that led to extirpation of the wolf in the Olympics three quarters of a century ago still stand as a major political hurdle to the animal's reintroduction.

In the rain forest valleys of Olympic National Park, it's possible to step into the profound beauty and ecological wholeness of a North America essentially unaltered by modern man. Here, wildlife populations still exist within the ancient cycles and dynamics of the forest, and processes of evolution continue uninterrupted. How much more stirring for us —and complete for the forest community — to have the haunting music of wolves restored to its place in this dynamic. It would add a singular note of wholeness to the persistent music of the rivers, and to the autumn whistling of the elk.

5
The Old-Growth Forest Community

It's a warm morning in late spring, and we've hiked deep into the old-growth forest of the Sol Duc Valley. We're looking for spotted owls, specifically a pair of owls known to inhabit this reach of forest. Park service biologist Erran Seaman has a map to the nest site, and he wants to know if the pair has hatched any young. We climb through an open forest of silver fir and hemlock. A half-hour after leaving the trail, at a little over 3,000 feet in elevation, we spread out and begin to traverse the slope. Within minutes Seaman finds the nest tree, a broken-off hemlock snag about three and a half feet in diameter. We're still unpacking equipment and setting up a tripod when the owl swoops down over our heads and lands on a nearby limb. From this distance, the northern spotted owl is a striking bird. A stippled brown and white breast fluffs out beneath a dark oval face ring surrounding soft gray feathered brows and alert dark eyes. The owl's interest is piqued as Seaman places a small mouse on a limb, but the owl remains aloof on its perch. It's nearly 15 minutes later, when I turn to ask Seaman a question, that the owl sweeps silently down, grasps the mouse in its talons, and flies to a perch nearer the nest. From there it whistles a contact call and hoots to its mate, a muffled *who-who-*

Focus of a contentious debate over the fate of remaining old-growth forests in the Pacific Northwest, the northern spotted owl is listed as threatened under the Endangered Species Act. Between 250 and 300 pairs nest in Olympic forests.

Janis Burger

who-who, with the mouse still in its beak. She answers, and the owl disappears into the nest cavity and reappears seconds later, *sans* mouse.

"A good sign," Seaman tells me. "The female is probably with her young." The previous year proved a poor one for owl reproduction on the Olympic Peninsula, and Seaman hopes this breeding season will show improvement. Throughout the summer, he and his crew will monitor some 42 owl pairs in the park, counting their offspring and determining their reproductive success. It's estimated that between 250 and 300 spotted owl pairs inhabit the 6,200 square miles of the Olympic Peninsula.

In 1990, the northern spotted owl was listed as threatened throughout its range by the U.S. Fish and Wildlife Service. Since the peninsula lies near the northern extent of the owl's range, and the population here is isolated from owls in the Cascade Range to the east, there is concern over the spotted owl's long-range viability in the Olympics. But it's not just the health of this singular species that fuels concern, it's the health of the old-growth forest ecosystem they've come to symbolize.

Spotted owls have been thrust onto center stage in an ongoing controversy over the fate of remaining old-growth forests. Estimates vary, but it's commonly agreed that less than 15 percent of the original old-growth forest remains in the Pacific Northwest. Spotted owls need old-growth forests for nesting and feeding habitat. That's why forest service biologists chose the northern spotted owl *(Strix occidentalis caurina)* as an "indicator species" for the entire old-growth ecosystem. Plucked from the obscurity of its deep forest home, the spotted owl now serves as a kind of canary in the coal mine. By monitoring the owl's population trends, biologists attempt to assess the health of the old-growth community.

The largest preserve of temperate old-growth forest in the Northwest, Olympic National Park offers scientists an opportunity to study an undisturbed old-growth ecosystem. Owl studies and other old-growth research conducted here may help with the future recovery and restoration of previously logged forests. More than 430 species of birds, mammals, amphibians, reptiles, and fish are known to inhabit the Olympic Peninsula as permanent or seasonal residents (another 50-odd bird species appear from time to time). All rely to some extent on the healthy functioning of this diverse forest ecosystem. For now, the spotlight shines most glaringly on spotted owls.

Owls, Squirrels, and Truffles

Nesting in cavities near the broken tops of old-growth trees, spotted owls can range great distances in search of their favored prey, northern flying squirrels and bushy-tailed woodrats. Both prey species are much less abundant in the Olympics than in coastal forests farther to the south, which may account for the larger home ranges of peninsula owls — from 3,000 to 33,000 acres per pair. Spotted owls can maneuver deftly in the winding canyons and multistoried layers of the old-growth forest canopy; the same canopy hinders the larger great horned owls that prey on spotted owls. Flying squirrels, which make up most of spotted owls' prey on the peninsula, also prefer the forest canopy, except when they're drawn to the forest floor to feed on truffles. The spotted owl–flying squirrel–truffle story is a perfect example not only of the kinds of specialized interactions that have evolved in these ancient communities, but of the complex web of mutually beneficial interconnections between species that form the heart of this remarkable ecosystem.

It's somehow appropriate that spotted owls, one of the most prominent members of the old-growth community, should be intimately connected with one of the forest's most hidden fruits. Truffles are the fruiting bodies of certain fungi. Unlike their above-ground cousins the mushrooms, truffles grow just beneath the surface. Fungi are everywhere in these forests; their slender, nearly invisible mycelia permeate and cover every vertical foot of it, from canopy needles to root hairs. In the canopy, fungi join algae, yeasts, and bacteria to form a microscopic community on and within the needles that scientists call simply "scuzz." Some of these fungi are toxic to needle-grazing insects, others fix nitrogen from the atmosphere, while others serve as food for a number of arthropods such as mites and springtails — as well as spiders and other predators. These predators in turn keep other leaf-eating insects in check.

Fungi also play a critical role in forest soils, where they form mycorrhizal relationships with the roots of trees. The mycelia, the vegetative, nonfruiting parts of these fungi, grow on and penetrate root tissues. Their threadlike hyphae, which are about one hundredth the diameter of a typical root hair, easily penetrate the interstices between soil particles and greatly enhance a root's ability to absorb moisture and nutrients. These fungi house nitrogen-fixing bacteria beneficial to both fungus and host tree. They also stimulate root growth and inhibit harmful pathogens and root grazers. For their part, trees feed sugars and amino acids — products of photosynthesis — to the fungi, which cannot photosynthesize. Mycelia networks also allow trees to exchange carbon and nutrients among themselves, a wonderful function that allows understory trees to benefit from the photosynthesis of larger trees in the canopy above them. Most plant roots are colonized in this way, including all Northwest trees. Foresters have found that in the absence of fungi, trees do not grow as well, and many forest seedlings will not survive. Fungi play an equally important role in cycling nutrients back to the forest from dead and decaying wood and plant matter.

Mushrooms are familiar to all of us, truffles less so. But their pungent aromas are well known to forest creatures and are irresistible to a number of small mammals. In the old-growth forests of the Olympics these include Douglas squirrels, Townsend's chipmunks, deer mice, red-backed voles, and flying squirrels — not to mention elk and deer. Fungal spores survive in the intestines of many small mammals, where they combine with bacteria and yeast, which makes them easily used by tree roots. The fecal pellets of small mammals scurrying about the forest are thus an important means of distributing these spores, especially in newly disturbed areas, where they pave the way for new tree growth. Although flying squirrels spend most of their time in the forest canopy, gliding deftly from tree to tree and feeding on lichens, the lure of truffles draws them frequently to the forest floor, especially during the spring to fall fruiting season. There, busily digging into the soil for these tasty treats, they become easy targets for spotted owls swooping down on silent wings.

By closely monitoring spotted owl populations in the Olympics, biologists like Seaman keep a finger on the pulse of the forest ecosystem. For it's not only the species of the old-growth forest, but the vital processes and interrelationships that have evolved here that hold the magic of this ancient realm.

The Architecture of an Old-Growth Forest

"Old growth" is a condition that develops over time in the coniferous forests of the Northwest. The major elements of the old-growth community were in place here when the climate stabilized about 5,000 years ago, but the finely tuned relationships between species and their environments suggest associations that reach back much further than that, possibly through the climatic fluctuations of the Pleistocene. It may be that the genetic wisdom encoded in the ancient forest ecosystem contains the ability to respond and adapt to dramatic shifts in climate and weather patterns we newcomers have yet to witness in the Northwest. We know

An old-growth forest of western hemlock and Douglas-fir (Doug-fir is easily recognized by its deeply furrowed bark). Unlike Douglas-fir, hemlock can seed in and thrive in the shade of an existing forest. In the unlikely absence of natural disturbances such as windstorms and wildfires, western hemlock will eventually dominate a stand.

Janis Burger

the old-growth ecosystem has thrived through repeated episodes of wildfire, wind, volcanic eruption, and glacial advance. Whether it can survive the onslaught of human design remains a troubling question.

It takes between 175 and 250 years for old-growth characteristics to begin to appear in most forest stands in the Northwest (that's at least a century longer than cutting rotations on most industrial tree farms). A 300-year-old stand is still "young" for old growth. Large trees are certainly an important part of an old-growth forest, but so are large standing snags and downed trees in various stages of decomposition. The uneven, partly open canopy that results when old trees fall and understory trees reach to fill the void are part of the old-growth story, as are the sunny glades that usher in a diversity of plants during the often decades-long interim. As is dramatically evident in the rain forest, an abundance of epiphytic plants, lichens, liverworts, and mosses also characterizes this unique forest community.

These conditions, combined with the moderate climate (cool in summer, warm in winter) and the long-term stability found within these centuries-old forests, have created a specialized habitat that nurtures a richness of species and complexity of relationships extending well beyond the spotted owl. At least 70 species of birds, 30 small mammals, and 20 amphibians are known to be associated with old-growth forests. One researcher noted that 40 wildlife species that use old growth or mature forest for primary habitat cannot meet their habitat needs outside this forest type. Debate continues over the number of species that actually depend on old-growth forests as habitat. Some, such as pileated woodpeckers and northern goshawks, seem to do well in younger natural or unmanaged forests (which usually retain a biological legacy of old growth: snags, downed logs, a mix of species). Others, like marbled murrelets and Vaux's swifts, appear solely dependent on this forest type. Still others, such as the endemic Olympic torrent salamander and Cope's giant salamander, are only rarely found outside old-growth forests.

Lives of the Old-Growth Forest

Until quite recently the marbled murrelet *(Brachyramphus marmoratus)* remained the most mysterious of seabirds. The last of North America's breeding birds to have its nest discovered, ornithologists searched for the nests of this robin-sized seabird for most of this century. There were plenty of clues to their whereabouts, however. Murrelets were frequently seen flying inland from the ocean, sometimes carrying fish, and their calls were often heard as the birds passed over coastal forests. They were so common early in the century that loggers working in coastal forests dubbed them "fog larks." Chicks occasionally turned up on the floors of old-growth forests, but it wasn't until 1974 that a nest was finally discovered. It was in a depression on a moss-covered limb high up in an old-growth Douglas-fir. Since then, a number of nests have been found in coastal forests of the Olympic Peninsula and elsewhere, some as much as 35 miles inland from the coast.

All nests found south of Alaska have been located in the upper canopy of large old-growth trees. Murrelets choose broad, horizontal conifer limbs covered with a thick layer of epiphytic mosses and lichens for nest sites, usually hidden beneath the shelter of low overhanging foliage. Females lay a single egg in a shallow depression in the moss, and the parents take turns incubating. After the chick hatches, both parents make the lengthy ocean-to-forest commute to feed it, sometimes as often as four times a day.

When fledging time approaches, young murrelets pluck the down from the tips of their feathers, exercise their wings, and are ready to fly. The parents usually stop bringing food at this time. Remarkably (for young murrelets seem aerodynamically unintended for flight—one researcher likened them to flying potatoes), the hungry fledglings leap from their nest and fly directly to salt water, where they land and begin to fish.

The decline of marbled murrelets following the destruction of old-growth forests on the Pacific Coast was pronounced. Breeding populations are no longer found in areas where coastal old-growth forests have been logged, and it's estimated that around 20,000 birds remain along the California, Oregon, and Washington coast. In 1992, after years of foot-dragging, the U.S. Fish and Wildlife Service belatedly listed the marbled murrelet as threatened in those three states. Viable populations still exist in the northern parts of the species' range along the Gulf of Alaska where, in the absence of ancient forests, some birds nest on the ground. Elsewhere throughout their range, marbled murrelets, like the wild salmon that return to their natal streams beneath the forest trees to spawn, embody the unity of forest and sea.

Vaux's swifts also rely almost exclusively on old-growth forests. True to their names, these small streamlined birds are astounding fliers and spend most of their lives in flight, cruising the high forest canopy and devouring insects on the wing. They can fly as much as 600 miles in a single day. Vaux's swifts construct their nests in the large, hollow, often burned-out snags of old forests. They nest and roost in colonies, but unlike their eastern cousins the chimney swifts have not taken to chimneys for nest sites. Studies have shown that Vaux's swifts are much more abundant in old-growth than in younger forests, no doubt because of the availability of suitable snags.

Western redcedar grows throughout the lowland forest, usually favoring streamsides, seeps, and swamps. Native Northwest cultures found a myriad of uses for redcedar, and its clear, rot-resistant wood is still prized for roofing and siding. As a result, most old-growth cedar stands outside of the park have been logged.

Janis Burger

Another distinctive snag dweller common to peninsula forests is the pileated woodpecker. Largest of the woodpeckers after the ivory-billed (which disappeared and is almost certainly extinct after the destruction of old-growth forests in the East), pileateds sport flashy red crests and long chisellike beaks. They feed primarily on larval and adult carpenter ants, and the staccato drumbeats of their jackhammering on snags to get at them is unmistakable. Pileateds excavate roosting and nesting cavities in large snags two feet or more in girth. Roomy, dry, well insulated, and lined with wood chips, many of these cavities are later put to use by Douglas and flying squirrels, owls, bats, and martens for nesting and roost sites. Wood ducks and fishers also favor cavities in large snags, and

Douglas squirrels, or "chickarees," are common residents of most Olympic forests, nesting in trees and feeding mainly on conifer seeds. Chickarees can often be seen gathering and storing cones in late summer and fall. Like many rodents, they also have a taste for fungi and are important agents in disseminating spores.

Ruth Kirk

larger mammals such as bobcats and even bears are known to "hole up" in them. As snags age and soften with decomposition, they become easier to work. By then, some 35–40 percent of the bird species common to Northwest forests may have used them for nest sites. Raptors such as goshawks and sharp-shinned hawks favor snags for perches and for "plucking posts," where they pluck and eat their kills. A number of bats, including long-eared and long-legged myotis as well as silver-haired and hoary bats, find optimal breeding habitat beneath the loose sheaths of bark that clothe old Douglas-fir snags.

Though not dependent on the peninsula's old-growth forests, Douglas squirrels, or chickarees, are found throughout the Douglas-fir forest, and their territorial chatter punctuates almost any walk in the woods. Chickarees do reach their highest abundance in old-growth forests, however, and they're all but absent from clearcuts. Their busy, gray-brown shapes, brightened by orange underparts and eye-rings and often crowned with their bushy upturned tails, are a common sight in Olympic forests; the cone husks of their middens decorate many a downed log or stump. Autumn in the silver fir zone can be a particularly hazardous time for hikers, as cones cut by squirrels rocket to the forest floor like small bombs. Douglas-fir cones cached for the winter and forgotten have proved important seed sources following natural disturbances in the forest. Chickarees are also important dispersers of fungal spores throughout the forest. Because they harvest and store so much food, Douglas squirrels remain quite active through the winter, and they're not above spending precious energy harassing trespassers from the safety of a limb. Quick as they are, squirrels do serve as prey for owls, hawks, and martens, among others. But the presence of predators hasn't lessened their brashness, nor has it seemed to put a nick in their population.

Martens and fishers, both sleek and accomplished hunters, are members of the weasel clan. The marten, the smaller of the two, prefers montane forests where it hunts a number of small mammals relentlessly and effectively. Its yellow-brown coat, low rounded ears, and orange throat patch are unmistakable. Though martens can hunt in the canopy (they share with spotted owls a taste for flying squirrels), winter frequently brings them to the forest floor, where they search among the downed logs of old-growth forests for rodents. They also use logs, well insulated with snow, for winter resting spots.

Martens, or their tracks, are commonly seen in the Olympics, but not so fishers. Their sleek, dark, white-tipped fir is said to resemble Russian sable and fetched a handsome price earlier in the century. Thirty-seven fishers were trapped in the Queets Valley in 1920; another 20 animals were taken from the Quinault the following year. The last documented fisher on the peninsula was a female trapped near Lilliwaup Swamp in the eastern Olympics in 1969. Fishers prefer to hunt in old-growth forests of the lower valleys; they flounder in the deep snow at higher elevations. Unfortunately, these are the most heavily logged forests on the peninsula. Concern over their numbers led forest and park service biologists to initiate population surveys in the mid-1980s, but they failed to turn up a single fisher. Though a number of sightings have been reported on the peninsula in the last 50 years, biologists fear that fishers may be threatened here and throughout the state.

Along with these more conspicuous species are smaller animals of Olympic's ancient forests that visitors rarely see. The brown creeper, an inconspicuous forest bird, blends so well with the bark under which it nests that it can evade predators simply by freezing in place. Townsend's warblers hawk for insects all summer in the high forest canopy where they nest, and frequently linger into fall and early winter. Golden-crowned kinglets inhabit the forest year-round. Hidden in downed logs and in burrows beneath the ground, redback voles dine almost exclusively on truffles, while several shrews rustle about busily in search of beetles and other invertebrates. Even the ubiquitous deer mouse shows higher numbers in old growth than in younger forests. These and countless other species are as integral a part of the old-growth ecosystem as spotted owls or Roosevelt elk.

The Lowland Forest

Whether hiking the trails or driving the Olympic Highway, travelers heading east from the rain forest valleys of the western Olympics to the drier northern and eastern slopes of the peninsula will notice the lowland forest takes on a distinctly different flavor. The western hemlock and Douglas-fir forest of the northern and eastern Olympics lacks the rain forest's lush extravagance of mosses, ferns, and epiphytes, for instance, as well as its abundance of nurse logs and resulting colonnades. East of the Sol Duc River, the domed crowns and deep green needles of grand fir replace Sitka spruce in the lower valleys and floodplains, and pines become noticeable throughout the forest. Large western redcedars still grace moist areas and stream banks, places where they have escaped the frequent wildfires, but the crowning glory of the lowland forest is the Douglas-fir.

First described in 1791 by Archibald Menzies, the naturalist who sailed with Captain Vancouver, it was more than a century later that David Douglas collected the seeds of this tree and sent them back to England. Not a true fir at all, it was known variously as a pine, a "yew-leafed fir," and a spruce. Taxonomists finally settled on "false-hemlock" (*Pseudotsuga menziesii*), or as it's commonly known, Douglas-fir. Simply stated, Douglas-fir is a magnificent tree. Found in a number of different forest types throughout the West, it reaches its greatest size in the old-growth forests of the coastal Northwest.

The deeply furrowed, rust brown, often fire-charred bark of ancient Doug-firs always catches my eye in the shade of a mixed forest. As early morning or evening light threads its way through the columns of the forest, an old Douglas-fir can seem as luminous as a softly glowing lamp. Its coarse-textured, moss- and lichen-flaked bark seems to invite the hand, and a few old giants alongside popular trails show evidence of years of "hugging."

Olympic National Park harbors some of the finest stands of old-growth Douglas-fir remaining in the Pacific Northwest. Outstanding examples can be found at Barnes Creek on the shores of Lake Crescent, along the Sol Duc Road and the short trail to Sol Duc Falls, and throughout much of the Elwha Valley. The upper end of this valley, where the Elwha begins to bend west above the Godkin River, supports an exceptionally beautiful forest with trees reaching diameters of eight feet or more. Protected in its valley bottom enclave in the heart of the mountains, some of these giants may date to wildfires that burned through the area in the early 1300s. The Skokomish River valley in the southeast corner of the park also supports stunning Douglas-fir forests; some may be easily seen from the short Shady Lane Trail at Staircase. A magnificent Douglas-fir grove on forest service land can be viewed along the Quinault Nature Trail on the south shore of Quinault Lake.

Taking in one of these grand old Douglas-firs from base to crown never fails to lift my spirit; the first stout limb emerges a hundred feet up the trunk and the whorl of limbs beyond fades into the splintered green light of the canopy. Looking out over a valley or mountain slope thick with the broken crowns of ancient firs suggests that all is right with at least one corner of the world, so there must be hope for the rest.

Though it is among the longest-lived trees of the Northwest, capable of reaching 1,000 to 1,200 years in age, Douglas-fir is not the climax tree species of the lowland forest. It can successfully reproduce only after a major disturbance clears standing forests and exposes mineral soils. Once established, though, it is vigorous and quick to grow, outcompeting other tree species and dominating young stands. Individual trees can eventually reach heights greater than 250 feet and diameters of over eight feet. The clear, strong, straight-grained wood of old-growth Doug-fir built the timber industry in the Northwest, and it remains the prime commercial timber species on the world market today.

Since Douglas-firs are unable to reproduce in dense shade, western hemlock is considered the climax tree species of the lowland forest. In the absence of large disturbances, a Douglas-fir forest will be slowly replaced by shade-tolerant hemlocks and western redcedars. But this textbook version of forest succession actually occurs only in the wetter parts of the forest. Elsewhere, the lowland forest proves itself a much more dynamic system. Not only do catastrophic wildfires and windstorms "recycle" forest stands, but floods, mass slumping, ice storms, diseases, and insect infestations also take their toll. Even where forests are left undisturbed, long-lived Douglas-firs continue to loom over the younger forest like proctors in a schoolyard. In the rainshadow of the northeastern Olympics, where the fire frequency is compounded by scarce rainfall, Douglas-firs maintain climax communities on a few exposed south-facing sites, but such stands are rare, amounting to less than one percent of the forest.

Like the temperate rain forests of the western Olympics, rainshadow forest communities of the northeast are another essential part of the Olympic ecosystem. They represent an adaptation to a much harsher, drier environment — conditions common to the intermountain West, east of the Cascade crest. In the rainshadow of the Olympics one can encounter trees typical of southern Coast Range or Rocky Mountain forests, and a number of understory shrubs and plants thrive here that occur rarely, or not at all, in west-side forests. Oxalis, that shamrocklike ground cover that carpets forest floors of the western slope, disappears in the rainshadow. In its place, the tough, low-growing shrub salal and the spiky leaves of Oregon grape crowd the forest floor. Baldhip rose is much more abundant, and the prolific pink blossoms of Pacific rhododendron serve as the signature of the dry-side forest. The pale blossoms of Pacific dogwood and bitter cherry trees also add brightness the spring woods. Two members of the heath family typical of drier climates, the shrubs kinnickinnick and manzanita, are also common to rainshadow forests, as are the shrubs snowbrush ceanothus and ocean-spray.

A tree that stands out among the lowland forests of the rainshadow is a graceful hardwood, the Pacific madrona *(Arbutus menziesii)*. The only broadleaved evergreen tree found in peninsula forests, madrona retains its dark, shiny leaves in winter. Instead, it sheds its thin outer bark to reveal wonderfully sinuous tan to purplish brown limbs. Clusters of

One of the hidden jewels of the forest, the Calypso orchid, or fairy slipper, brightens mossy trailsides. Fairy slippers rely on the presence of certain fungi in the soil to successfully reproduce.

Janis Burger

pink-tinged flowers in spring yield bright orange-red berrylike fruits in autumn, making this one of the loveliest of woodland trees.

Such continental tree species as quaking aspen and Engelmann spruce put in rare appearances here, as does common juniper at higher elevations. A rare Rocky Mountain juniper woodland can be found in the upper Dungeness country, and dense stands of lodgepole pines reminiscent of the northern Rockies grace high south-facing slopes. Whitebark pines are a timberline tree in the open meadows of the upper Dungeness, just as they are along the high, windswept ridges north of Yellowstone.

The lower forest also follows patterns common east of the Cascades. In the lowlands of Dungeness Valley near Sequim, there is a second timberline. Rainfall drops below 15 inches a year, and forests give way to prairie. Growing among the sprawl of the present town of Sequim, the last remnants of a Garry oak woodland attest to an arid, presettlement savanna landscape similar to Oregon's Willamette Valley. Sequim Prairie, along with other extensive rainshadow prairies here and on nearby islands, may have been maintained by Native American burning for the past 3,000 to 4,000 years. Some 80 plants common to the Sequim Prairie are known to have been used by native people in the Northwest, and camas bulbs, found only in prairies, were a staple of Native Americans' diets. Some of the same species found in remnant prairies today also grow on rocky, south-facing slopes of the Dungeness uplands, suggesting the mountains here may have served as a biological refuge during the ice ages. Sequim Prairie was known to early settlers for its abundance of game animals and was hunted heavily. In some ways the Dungeness Valley was a small Serengeti on the northeast peninsula, a wonderfully rich assemblage of open habitats, grazing herbivores, and their predators, with small numbers of humans playing an active role in the overall ecology. Today, a few native prickly-pear cacti *(Opuntia fragilis)* still bloom in undisturbed corners of the Sequim Prairie, but agriculture and urban growth have altered the landscape irrevocably.

These unusual plant communities found throughout the northeast rainshadow of the Olympics are astounding for their proximity to one of the wettest spots in North America. More than a reflection of this area's low rainfall and active fire history, they are the botanical legacy of the region's long-term climatic history. As we face the possibility of global warming, these drought-tolerant, fire-adapted, rainshadow communities may serve as critical genetic reserves.

Fire and the Forest Community

Just as wind played a dominant role in forest disturbance in the peninsula's coastal forests, with major windstorms hitting the coast about every 20 years, wildfire has been the major catalyst shaping lowland and montane forests elsewhere in the peninsula. Before settlement, large naturally caused fires swept through the forests of the Olympic Peninsula countless times. Evidence of fires in the rings of old-growth trees shows at least three periods of major fire activity on the Olympics in the past 1,000 years.

In the year 1308, a large fire swept across millions of acres in the western Cascades and Olympics, burning about half the forests on the peninsula. The Douglas-firs that seeded in on the scorched soils in the wake of this fire are now 650-plus years old and can be found throughout the park. About 250 years later, another major conflagration hit the peninsula's forests. This fire swept through mountainsides and valley bottoms across the Olympics, often reburning areas recovering from earlier fire. Large expanses of forest in the Washington and Oregon Cascades also burned at this time.

The last great series of wildfires to hit the peninsula swept through on dry east winds around 300 years ago, in 1701. Those fires burned over a million acres of lowland forest along the eastern and northern slopes of the Olympics from tidewater to middle and upper elevations. The great lowland Douglas-fir forests that greeted early settlers and became the basis for the peninsula's early timber economy had their beginnings in the ashes of this fire. The late 19th and early 20th centuries saw a rash of smaller fires, largely the result of early land clearing and the reckless logging practices of the day. These fires burned parts of the northern and eastern peninsula as well.

Unlike clearcut logging, natural disturbances like wildfires and windstorms are selective in what they take. As the Yellowstone fires showed, wildfires commonly burn in patches, leaving ribbons and islands of standing green trees, even in the most heavily burned areas. The thick corky bark of older Douglas-firs insulates them from most light and moderate fires and some high-intensity fires as well. Surviving trees serve as seed sources for the new forest, further selecting for this fire-resistant trait. Moist areas along streams, draws, and wetlands are frequently spared, allowing redcedars and other moisture-loving forest plants to survive. Large downed logs that have absorbed water like big sponges provide cool, moist islands during fires where small mammals, amphibians, and invertebrates can survive. Downed logs also protect fungi needed to form mycorrhizal relationships with new seedlings. Fire-killed trees that remain standing serve important habitat needs for a number of snag-dependent wildlife species.

The forest has evolved with these disturbances, and they are essential in maintaining the ecosystem. The pattern of different-aged forests that results from wildfires overlays topographic differences in rainfall, elevation, and exposure to create a biologically rich mosaic of natural forests across the Olympic landscape.

The Montane Forest

Farther up the mountain slopes, beginning at about 1,000 feet in elevation in the western Olympics, higher east of the Bailey Range, the lowland forest grades into mountain forests of Pacific silver fir. Silver fir, or "lovely fir" *(Abies amabilis)* as it's known to science, is a tall, stately conifer with resinous silver-gray bark and deep green foliage. The fires that raged through much of the western hemlock–Douglas-fir forest over the last 700 years have largely missed this cool, wet forest zone, making silver fir–hemlock among the oldest forest communities (if not the oldest trees) in the Olympics. Western hemlocks of the lowlands persist upslope to become part of this zone, but silver fir's extreme tolerance of dense shade leads to its eventual domination of the montane forest. Conversely, silver firs can also be found at low elevations, particularly on north slopes and draws where pockets of cool air flow down from higher slopes. The floors of west-side silver fir forests are usually carpeted with damp beds of oxalis and moss. In the eastern Olympics, this forest type is confined to narrow bands along cool, mid-elevation stream bottoms, and huckleberry shrubs and five-leaved bramble grow below. In the dry heart of the rainshadow, the silver fir zone drops out completely. In its place, fire-generated forests of Douglas-fir and hemlock, mixed liberally with a few species of pine, juniper, and yew, extend all the way up to high windy stands of subalpine fir. Throughout the rest of the Olympics, the silver fir forest finds its upper limit in shaggy, snow-loving forests of mountain hemlock.

But anyone who spends time hiking in the Olympics knows that forest types really don't stack up that neatly. The way we group species reflects our human need to classify as much as it does natural processes. Mountain species like Alaska yellow cedar and lowland giants like Douglas-fir both appear throughout the silver fir forest, and any turn in the trail, any creek gorge can yield something new. During the years I worked the trails of the Olympics, I wondered at the large old Douglas-

Janis Burger

firs I'd find scattered throughout the upper reaches of the silver fir forest; so much larger and obviously older than the trees of the montane forest that surrounded them, they seemed out of place. It wasn't until years later that forest ecologist Jan Henderson unlocked the mystery for me. Henderson had spent a decade completing a comprehensive ecological survey of the Olympic National Forest. For him, the silver fir zone is a living legacy of the forest's response to 1,000 years of climatic change.

Montane forests of silver fir occupy a wide band between 1,000 and 3,000 feet in the western Olympics but are mostly confined to cool, moist stream bottoms on the east side. Pendulous lichens that cover these trees down to the average winter snow depth are an important seasonal food for flying squirrels.

Henderson explained that those venerable 700- to 900-year-old Douglas-firs scattered high in the silver fir forest are the last remnants of an earlier climate. Douglas-fir forests seeded in following large fires that marked the end of a thousand-year-long warming period. During that time the Douglas-fir forest had extended its range far up the mountain slopes, and the silver fir zone was a narrow band above it; a similar zonation exists in the Oregon Cascades today. Beginning around 700 years ago, with the onset of the global cooling trend known as the Little Ice Age, shade-tolerant silver firs worked their way downslope in a series of steps, eventually replacing the older Doug-fir forests. The silver fir forest took its last downward step between 150 and 200 years ago, just as the Little Ice Age was coming to a close. These montane "pioneers" that ventured into the lower forest aren't faring well under current climatic conditions. Succumbing to disease and insect infestation, they experienced serious die-offs in the 1950s. Today, the survivors describe a ragged line that marks the tail end of the Little Ice Age. Like the ancient Doug-firs above them, these silver firs bear witness to that elegant dynamic by which an ecosystem responds to climatic and environmental change.

Olympic harbors a wealth of pristine forest lands unequaled in any other national park. Nearly a quarter million acres, a quarter of the park's land, is crowned with the productive, diverse, and habitat-rich old-growth forests. These forest communities extend up river valleys on all sides of the park, weaving the biological wealth of the lowlands into the mountainous heart of the ecosystem. Outside Olympic National Park, logging and human-caused fires have destroyed more than 95 percent of the old-growth Sitka spruce and Douglas-fir forests that once crowned peninsula lowlands. The valley forests that remain on park, national forest, and state-owned lands constitute a resource of planetary significance. Stitched together by the rivers that link the mountains to the sea, they form the heart of this coastal ecosystem. Within them, an extended family of spotted owls, flying squirrels, elk, salmon, and humans share a relationship far older than the oldest of trees.

6
The Lives of Olympic Rivers

Of all the seasonal wildlife movements that weave together the elements of the Olympic ecosystem, the fall and winter migrations of salmon up the forested rivers must be the most dramatic. Salmon climb nearly every stream on the peninsula, but to me the ascent of coho salmon into the deep mountainous reaches of the upper Sol Duc River is surely the most spectacular. Gold-brown needles litter the forest floor, and high winds usher in the first fall storms. For days rains have saturated mountain slopes, and tributary streams pulse with runoff. The river has risen in its bed, sweeping the summer's litter from gravel bars and burying boulders beneath cascading blue-green waves. It seems an unlikely time for the ocean to bestow gifts on the inland forest; the headlong rush in the valleys is seaward. Falls and rapids churn, and low limbs along the rivers bob and tug in the runoff. But a visit to Salmon Cascades, a short falls where the Sol Duc River narrows and spills over a tough shoulder of basalt, brings me to the heart of this yearly exchange. With the coming of the fall rains the cascades have erupted into a whitewater torrent. For weeks, a run of coho salmon that entered the river in August has been working its way up the low boulder-strewn channel and has stacked up beneath the falls. Now, the "summer-run" coho take flight.

As I lean against the wet rail of the overlook beneath dripping trees, first one, then another, then three salmon at once leap from a sudsing pool at the base of the falls. Kicking wildly, they fling themselves in arcs above the white rush of the torrent. Their dark flashing shapes, lit with a red underglow, seem to float almost motionless before falling and being swept back downstream. Above the falls in calm, spring-fed pools beneath a forest of Douglas-fir and redcedar lie the spawning gravels of their birth. It seems, amid the tumult of falling water, that few if any will make it past this barrier. But with thousands of years of adaptation to the dynamics of this very stream behind them, nearly all of them will. With them, the salmon bring a wealth of nitrates, phosphates, and other nutrients from the ocean to the forest.

No one knows why the summer-run coho enter the river so much earlier or climb farther upstream than other wild coho. Coho salmon (*Oncorhynchus kisutch*) are common in the Olympics, where they spawn in small streams, tributaries, and the side channels of most major rivers. But by entering the mouth of the Quillayute River during the low flows of August, the salmon are forced to contend with the shallow rocky flow of the river for 50 miles inland. Along the length of their route they are preyed upon by black bears, bald eagles, and those sleek, efficient hunters of the waterways, river otters. Like others of their kind, they have just spent two years in the vast waters of the north Pacific, feeding and growing strong on its plentiful resources: smelt, herring, rockfish, and sand lance, as well as smaller salmon. By eating voraciously during their final spring and summer they have doubled or tripled their weight and entered the river at peak physical condition. Now, not having fed for as long as two months in the river, they must surmount this final barrier and still maintain the strength to spawn. It is a cycle repeated throughout the rippling waters of peninsula rivers, a ritual both sacrificial and heroic.

Upon reaching their spawning grounds, females excavate their nests, or redds, in clean washed gravel with great kicking sweeps of their tails. Males contend for each available female, lunging and biting with hooked snouts and jagged teeth, emblems of their spawning stage. When a female is ready to spawn, a dominant male approaches and swims close beside her. Seen from above, they articulate a single graceful motion. Crouching low in the redd, the female releases her eggs and the

Janis Burger

male his milt simultaneously. Then the female moves just
upstream and covers the eggs with gravel from her next
redd. Spawning gravels must be large enough to allow a
flow of cold, oxygenated water to bathe the eggs during
the three to five months of incubation and protect them
from predators as well as heavy flows. As many as 5,000
eggs may be laid in several redds by a single spawner.

*With the first heavy fall
rains, coho salmon
begin to leap the torrent
of Salmon Cascades
on the Sol Duc River.
Once past the cascades,
they spawn beneath
old-growth forest up-
stream from Sol Duc
Hot Springs.*

The salmon are battered and weakened by their long
inland journey. White fungus covers the scars of their
upstream passage, and fins and tails are tattered and frayed. Their last
reserves of ocean-gleaned energy nearly spent, females guard their redds
while males summon strength to spar for the last available females.

Within days the ritual is complete, and the salmon die. Carcasses litter gravel bars and log jams and catch on the brush that borders the
stream. A pungent smell of decaying fish hangs in the chill air. Beneath
the gravel, encoded within the soft pink quarter-inch eggs, is the genetic
information that will insure that enough of the spring fry survive their
first perilous year in the stream and two more in schools at sea to successfully navigate the river and spawn in these same sheltered waters.

Olympic is the only national park outside Alaska that supports such generous runs of wild anadromous fish. Throughout the peninsula's rivers, tributaries, and small coastal streams, numerous stocks of the five species of Pacific salmon native to the Northwest as well as steelhead trout repeat this cycle. A stock is a genetically distinct population of fish that spawn together in a single stream. Like the summer-run coho of the Sol Duc, each stock is perfectly keyed to the geological character, climatic variables, disturbance history, and flow regimes of its native stream. Wild salmon cycle from river to ocean and back to stream, carrying in their genetic makeup the evolutionary wisdom of their watersheds. The clean, cold, forested rivers of Olympic National Park provide some of the most productive spawning habitat on the Northwest coast. Salmon and steelhead stocks native to the Olympics represent an irreplaceable resource as much a part of this islandlike ecosystem as the old-growth forests in which they evolved.

To me, salmon embody the essential unity of mountain, forest, and sea. In the downstream flow of minerals and nutrients, they are agents of renewal and return. The evolution of Pacific salmon, their adaptation to the postglacial streams — and the forests that lent them structure and stability — is a miracle of reciprocity. A timeless cycle of death and rebirth informs their migrations, and the beauty of these seafaring creatures resonates through the forests, rivers, and streams like a slow and silvery pulse.

Wild Olympic Salmon

Rarest of the peninsula's salmon is the sockeye. Adapted to lake environments, juvenile sockeye emerge from the gravel and spend a year in coastal lakes before migrating to the ocean for another one to four years. On their return they often summer in lakes before spawning. The sockeye has been badly hurt by hydroelectric dam construction in the Northwest; it's estimated that 96 percent of sockeye habitat in the Columbia River basin alone has been lost. Only a few wild runs are left in Washington state, and two occur in the park. A large run spawns in Quinault Lake and its tributaries in late winter and early spring. In spring and early summer a much smaller run enters Ozette Lake on the park's northern coastal strip and spawns in its tributaries in winter. My fishing friends consider the dark rich flesh of wild sockeye without peer.

Largest of the wild swimmers, Chinook or king salmon spawn in cobble-sized gravels in the main channels of the peninsula's large rivers. Small spring runs occur in the Hoh, Queets, and Quinault rivers, where they favor the headwater reaches (swimming as much as 45 miles up the Hoh), as well as the Gray Wolf River in the northeastern Olympics. Summer runs also enter the Hoh and Queets rivers, as well as the Quillayute system, and spawn in the middle reaches. Both spring and summer Chinook remain in the rivers throughout the summer, where they're prized by sport fishermen. The majority of Chinook enter rivers in fall and spawn near the coast. Fall Chinook are found in all the park's coastal rivers.

Because coho or silver salmon (all salmon have at least two names) spawn in tributaries, spring creeks, side channels, and the upper reaches of main river channels, they are found virtually everywhere on the peninsula. But this ubiquity subjects them to some severely degraded habitats outside the park. Small streams, wetlands, and side channels are critical habitats for coho, but they are also among the peninsula's most heavily roaded, logged, and developed habitats. And since coho spend their first year in streams before migrating to the ocean, they are also susceptible to both winter floods and summer droughts. As a result, coho runs are depressed in many areas of Puget Sound, the Strait of Juan de Fuca, and the southern Washington coast. With prime coho habitats like the upper Sol Duc, the park protects some of the best remaining spawning and rearing habitat remaining on the Northwest coast.

Ironically, the two most abundant Pacific salmon, chum and pink salmon, are uncommon on the northern Washington coast (though chum are relatively abundant in Grays Harbor and Willapa Bay to the south). Pinks return to Olympic streams in the fall of odd-numbered years and spawn in the lower rivers, where their eggs are in danger of being washed away by floods. This problem is compounded in watersheds that have been heavily logged. A unique pink salmon stock reminiscent of the Sol Duc coho enters the Dungeness River system in early summer and climbs well up the Gray Wolf to spawn, some reaching the park. Small numbers of chum salmon spawn in the Queets, Ozette, and Quillayute rivers as well as some lower east-side streams, but it's believed they were never plentiful on the peninsula.

On the other hand, the peninsula has become justly famous among sport fishermen for its winter runs of wild steelhead, or sea-run rainbow trout. Steelhead spend an average of two years in fresh water and two years at sea. Unlike salmon, they can return to spawn two or three times. Steelhead grow large, up to 30 pounds in coastal rivers, and though smaller than Chinook salmon (which can reach 70 pounds), they are vigorous fighters and have a reputation for being notoriously difficult to land. Summer-run steelhead have declined precipitously throughout the Pacific coast, but winter stocks still lure fishermen hip-deep into the icy waters of coastal rivers to go nose to nose with "steelies."

While winter steelhead runs remain relatively healthy on coastal rivers, steelhead in east Olympic streams are in trouble. A 1992 report by the Washington Department of Fish and Wildlife and the Western Washington Treaty Indian Tribes lists winter steelhead stocks as depressed on the Dungeness, Dosewallips, Duckabush, and Skokomish rivers. The greater portions of all of these east-side watersheds lie outside Olympic National Park, where they have been subject to intensive road building, logging, and development. During heavy rains, roads dump silt and pollutants into streams; logging removes large trees and shade for streams, and residential development introduces wildcards into the streams—from septic systems, automobiles, and lawn and garden chemicals. Summer-run steelhead as well as coho salmon are also faring poorly on the Dungeness River, which is tapped for irrigation for farms in the arid Sequim Prairie. Dungeness Chinook and pink salmon are also listed as critical. Chinook eggs have been captured and reared in a hatchery to insure their survival. In 1999, Puget Sound Chinook, Hood Canal summer chum, and Lake Ozette sockeye were listed as threatened under the Endangered Species Act.

Streams that drain the rainshadow of the Olympics are subject to chronic low flows during dry summer months, making them extremely susceptible to degradation from shoddy land use practices. In fact, five of the twelve salmon stocks listed as critical in Washington state are found in the northeastern Olympics. So are a high percentage of depressed stocks, including pink salmon in the Dosewallips and coho in the Duckabush rivers. These and other eastern Olympic salmon runs are important precisely because they have adapted to these stressed environments. One biologist compares them to the more than 400,000 varieties of plants our government invests tens of millions of dollars in each year to conserve because of their genetic importance to commercial agriculture. The difference is that wild salmon can't be housed in a climate-controlled warehouse; to save their genetic diversity it's imperative to protect their habitats.

Degraded habitat isn't the only threat facing Olympic's wild salmon stocks, however. Even runs that spawn entirely within the park can be subject to overfishing at sea. This becomes a major threat to wild salmon when hatchery fish are figured into the allowable catch. The plight of the wild coho that leap Salmon Cascades illustrates the problem.

The threats hatchery fish pose to wild salmon are now well understood. Large numbers of hatchery-reared fry released into streams tend to school up, as if waiting to be fed. They can't read territorial cues, and displace native fry from feeding stations. They also compete with wild fish for resources at sea. Bred in close quarters, hatchery fish have also been known to carry diseases that can infect native salmon. But a larger problem develops when hatchery fish return to river mouths along with wild fish. Washington state officials commonly set commercial fishing limits based on the expected hatchery return, which is often many times greater than the wild return. This was the case with the Sol Duc coho. Of the 20,000 coho that returned to the Sol Duc in 1980, for example, only 1,800 were wild fish. But they were fished just as heavily as the hatchery fish offshore and in the Quillayute River. The state has since attempted to address this problem by cutting back on the allowable catch, but hatcheries remain a serious threat to depressed wild stocks, particularly in the smaller rivers of the eastern Olympics.

Taken together, the wild runs of salmon and steelhead that return to Olympic rivers each year represent an incalculable resource, not just to tribal, commercial, and sport fishermen, but to the terrestrial ecosystems that receive their bounty and to the productivity of the streams themselves. No one knows the size or extent of the historic wild runs that once thronged Olympic rivers, but they must have been staggering. Homesteaders reported salmon choking small streams on the peninsula, and tales of early travelers crossing streams on the backs of salmon, however unlikely, are still repeated. The historical record gives a more accurate glimpse.

Records for canneries operating on the Hoh, Queets, and Quinault rivers indicate that during the years 1911 to 1927, canneries processed an average of 76,500 salmon and steelhead per year — and more than double that number in the best years. Calculations based on similar records for Oregon canneries during that time suggest that a million coho salmon alone returned to Oregon coastal streams each year. Aside from its value to humans, this kind of yearly return represents a virtual bonanza of nutrients to freshwater aquatic and terrestrial ecosystems.

Salmon and the Forest Community

Concern over the decline in historic salmon runs in the Northwest and the implications for forest and wildlife communities on the peninsula prompted researchers from Olympic National Park and the Washington Department of Natural Resources to embark on a groundbreaking study. They wanted to see how and to what extent spawned salmon carcasses were used in the upstream reaches of spawning streams. Their findings shed new light on the ecological ties that bind the mountains to the sea.

The researchers distributed nearly 1,000 tagged coho salmon carcasses on sections of spawning streams in the Olympics, including the upper Sol Duc. A number were fitted with small radio transmitters. Movements of the salmon, both within and outside the stream were carefully noted. Since the tributary streams used by coho are subject to frequent flooding in fall and early winter, it was assumed that most spawned salmon carcasses were flushed downstream. Researchers found this was not so. A large percentage of carcasses remained in the spawning areas. They were caught in logjams and in the limbs of submerged logs, or stranded in pools and eddies formed by fallen logs. A large number were dragged by animals into the adjoining forest and partly eaten.

One of the study's most surprising findings was how many of the carcasses were consumed by wildlife. On the upper Sol Duc, for instance, the figure was 8 out of 10. Most were eaten by black bears, but no fewer than 21 other wildlife species were observed feeding on the salmon as well. Bears, river otters, and raccoons dragged the carcasses from pools, ate part of them and left the rest. Over the days and weeks that followed, a company of smaller mammals and birds tidied up. Among the guests at the banquet were coyotes, minks, skunks, bobcats, squirrels, deer mice, and several shrews. Bald eagles, red-tailed hawks, ravens, and crows

Data collected from coho salmon carcasses distributed in Olympic streams help scientists understand how spawned salmon contribute to the forest ecosystem. A researcher notes the location and condition of a carcass fitted with a radio transmitter to trace its movement by scavengers.

Janis Burger

flocked to river corridors to join them, as did Steller's and gray jays. Dippers, those small riverine wrens who usually dine on invertebrates and fry, couldn't resist the spread, nor could those shy, diminutive singers of the underbrush, the winter wrens. In all, just over half the total species of mammals and birds inventoried along the streams partook of the spent salmon. Only the mole, beaver, cougar, and elk abstained. Other studies have indicated that even deer may venture an occasional nibble.

This seasonal largess comes when other sources of protein may be scarce, and it immediately precedes a time of extreme stress for many wildlife species in the Olympics. Spawned salmon may make the difference between death and survival for some animals during severe winters.

It's extremely likely, with historical spawning levels of coho estimated at 10 times current numbers, that past runs supported much higher levels of wildlife than are found in parts of the Olympics today.

Virtually all members of the aquatic community benefit directly from spawning coho salmon — from aquatic plants and invertebrates to resident fish. Only the "shredders" (snails, stonefly nymphs, and other insects that break coarse organic material down into finer particles) are not enriched by salmon-derived nutrients. Growth rates for young coho salmon double in the winter following a spawning, while cutthroat trout in nearby streams without spawning salmon show no winter growth at all. Salmon add nitrogen and carbon to the whole spectrum of life in a stream and increase aquatic diversity. Not surprisingly, the elimination or reduction in number of spawning coho salmon in small coastal streams puts the entire stream community at risk.

The benefits of spawned salmon to streamside forests are also substantial. Salmon carcasses not consumed decompose, enriching the streamside plant communities and aiding in the decomposition of the autumn litter of leaves and woody debris. As the season progresses, elk and blacktail deer take shelter in these riparian forests, and the early leaf growth in spring provides an important part of their seasonal food. In fact, the riparian forests that border Olympic streams serve essential functions in the life histories of a number of wildlife species. The clear cold waters, deep pools, and clean spawning gravels necessary for spawning salmon are to a large part dependent on riparian forests, as are many of the creatures that feed on spawned salmon. Some 87 percent of the birds, amphibians, reptiles, and mammals found in western Washington use these streamside zones during certain seasons or parts of their life cycles. In the maritime Olympics, where the landscape is laced with rivers and streams, the network of riparian areas along streams, lakes, and wetlands are among our most critical as well as productive habitats.

The Interplay of Forest and Stream

Most river valley roads and trails in the Olympics skirt riparian forests. Shady bands of red alder, black cottonwood, and bigleaf maple mix with a lower growth of vine maple, Scouler's willow, and cascara to form ribbonlike oases of deciduous growth in the conifer forests. Western red-cedar, western hemlock, and Sitka spruce also form an important part of mature riparian communities. As transition zones between aquatic and upland environments, riparian areas provide the three essentials of wildlife habitat: food, cover, and water. Varying degrees of moisture within these areas support a mixture of plant communities. Birds such as warblers, vireos, finches, tanagers, grosbeaks, and thrushes use riparian areas for nesting and foraging. Common and hooded mergansers nest in cavities here, and ospreys and bald eagles build nests in the broken tops of large streamside conifers. As we've seen in the rain forest, riparian forests provide seasonal migratory routes, as well as calving and fawning areas and cover for Roosevelt elk and blacktail deer. They serve as travel corridors for a host of bird and mammal species and as critical habitat for a number of amphibians including some endemic to the Olympic Peninsula.

Riparian forests are essential to the health of the watershed in other important ways as well. The leafy limbs of deciduous trees and shrubs shade streams, keeping water temperatures cool enough for salmon and trout. Overhanging brush and roots provide havens where salmon fry and fingerling trout can hide from predators. Leaf-dwelling insects and their droppings provide food for aquatic invertebrates and fish, and fallen leaves and wood debris add important nutrients to streams, feeding aquatic invertebrates — the grazers, shredders, and collectors — that form the basis of stream ecology. Tree roots stabilize stream banks, and streamside plants filter runoff in wet seasons, further protecting water quality. Historically, one of the most important functions of riparian forests within the Olympic ecosystem was to render the spare, scoured, post-Pleistocene landscape fit for salmon.

In the millennia that followed the retreat of the Cordilleran ice sheet, the rivers that drained the Olympic Mountains were steep, raw torrents carrying loads of sediments from deeply incised canyons to braided floodplains. Forests made inroads into these newly carved valleys during the climatic fluctuations that followed the Ice Age, reclaiming mineral soils and stabilizing landscapes. But it wasn't until the climate approached that of today and forests matured and developed old-growth characteristics some 6,000 years ago that salmon were able to spawn and rear in these quick mountain streams.

Janis Burger

Large glacier-fed westside rivers like the Hoh meander across wide floodplains, toppling trees and building new river bars. Logs and rootwads strewn by the river help regulate high flows and create habitat for fish and invertebrates.

Geologically, the peninsula's rivers are still quite young. Fed by the heavy precipitation of the coast, they are still subject to floods and landslides. As old-growth forests developed, large conifers fell across rivers and streams, slowing the water and adding structure and complexity to these mountain streams. Downed logs create pools, riffles, and glides. On larger rivers, logs stack up in logjams, creating multiple side channels and dissipating floodwaters onto floodplains, where the current is slowed by vegetation and drops much of its sediment load, conserving soils and nutrients in the watershed. Downed logs stabilize shorelines and create meanders with their deep eddies and gravel riffles. The overall effect of the old-growth forest is to render streams and rivers less efficient conduits for flowing water, but optimal habitats for salmon and trout. Each blockage creates places where salmon and trout can rest, feed, and hide; gravels trapped by downed logs serve as spawning areas for salmon as well as rearing areas for insect larvae that feed the fry. Downed wood retains salmon carcasses in the stream, and nutrients from spawned salmon feed the forest. The wild stocks of Pacific salmon could never have adapted to the peninsula's streams without the development of old-growth forests across the landscape. The continued health of wild salmon today remains utterly dependent on the health of the forested watersheds.

Lives of the River

Though fall and winter present a movable feast for wildlife along salmon streams, Olympic's rivers teem with life in all seasons. Spring brings its own rich assemblage of species to the swollen streams, often when the high country is still locked in winter. Nesting waterfowl are most notable, and singularly conspicuous among these are harlequin ducks.

Harlequins look as though they might have been painted by tribal fetishists. Breeding males are blue-gray and russet, streaked with splashes of brilliant white edged in black. When they arrive in the forested reaches of Olympic rivers in March or early April to breed, it's as if the village minstrels were let into the cathedral. For me, spring hasn't begun until I've encountered these elegant and tough little ducks in the steep, fast-moving rivers of the northern or eastern Olympics. Like salmon, they return to their native rivers at breeding time. Once there, they surf the rapids, plunging and diving among boulders and pools to scour the substrate for invertebrates, then climb atop mid-stream boulders to preen. The moss-capped rocks of snowmelt streams grow vibrant beneath them.

Harlequins seem drawn to rough water, whether the whitewater of their breeding streams or the dark currents and plunging surf along the seacoast where they spend the rest of their year. They are the only duck in North America that breeds in steep mountain streams, and they add their splashes of color to some of the grandest landscapes in the country. In the Olympics, harlequins nest in nooks and on knobs of streamside cliffs or among branches and litter on the ground along streams. A rare cavity nest (one of two reported thus far) was found recently in a snag up the Elwha River. Soon after incubation begins, the drakes head back downstream to salt water. Females and their hatched young remain in the rivers through most of the summer. More than once I've held my breath along a river bank as a startled female and her brood half flew, half surfed across a boiling rapid to arrive intact in a far eddy.

You might think, with the whitewater of the Olympics for a nursery, there's no environment that would cause young harlequins concern. Yet the harlequin duck is currently a candidate for the federal threatened and endangered species list. Harlequin populations on the East Coast never recovered from earlier hunting for trophies and feathers. Some West Coast areas have shown marked declines, but harlequins have historically been plentiful in the Olympics. Recent surveys of harlequins in the Olympics found they are scarce on the lower meandering portions of large west-side rivers, but numbers increased as river gradients steepened. Highest harlequin counts were reported for the Elwha River, at more than two nesting pairs per river mile. In 1991, a minimum of 152 breeding pairs were estimated for Olympic streams. Only about a third of available females have broods in any given year, and the statewide population is estimated at around 1,400. In winter, harlequins from throughout the intermountain West converge on the outer coast and the Strait of Juan de Fuca to winter by the thousands.

If harlequins have any competition in their specialized breeding habitat, it can only come from that most intrepid of songbirds, the dipper. Dippers also breed along quick mountain streams, but unlike harlequins, they live there year-round. Easily recognized by their plump slate-gray plumage and constant bobbing motion while on boulders or on shore, their song carries above the noise of the loudest rapids and ranks among the sweetest of mountain sounds. John Muir called dippers "the hummingbirds of blooming waters," and wrote that they love "rocky ripple-slopes and sheets of foam as a bee loves flowers." Alone among songbirds, dippers have forsaken foraging in forests and fields; instead they feed in the roiling waters of the mountains. Diving under water, swimming and walking along the rocky substrate, they work the gravels for insect larvae like trout. Upon emerging they fly upriver at spray height or briefly rest, "dipping" on a rock.

Ornithologists disagree over the cause of this bobbing behavior; some suspect it's a camouflage mechanism effective in their quick-moving environment. I've often suspected it might have more to do with keeping warm in the damp shade of river canyons, but their thick plumage and oversized preen glands for waterproofing their feathers seem to render them impervious to the icy waters. And their spring singing is undaunted even when neighboring shorelines are still thick with snow.

Commonly seen (and heard) along streams, lakeshores, and protected coastal areas on the peninsula, belted kingfishers are superb divers and expert fishers. They nest in tunnels dug in bluffs and river banks and keep watch for fish from overhanging limbs along streams.

Janis Burger

Dippers' large mossy nests are built along cliffs, often behind water-falls; in downstream reaches, bridge abutments will suffice. Both adults tend the young. After their mating season, dippers follow the thawing drainages upstream. Later in summer, they can be seen working the edges of high mountain lakes. In winter, dippers migrate downstream to lower rivers or large lowland lakes. Like their smaller (and equally melodic) cousins the winter wrens, dippers lend spirit and warmth to river walks, even in deepest winter.

Another conspicuous bird common to Olympic streams is the belted kingfisher. The loud rattling calls of kingfishers are unmistakable and are usually heard before the bird shows itself along the waterway. Large-headed, with a long stout beak, a kingfisher's compact body seems expertly designed for diving and catching fish, a feat it accomplishes with unfaltering accuracy. Kingfishers visit virtually every Olympic stream as well as the shores of the strait and Puget Sound. In salt water I've watched kingfishers dive from a hover, but on streams they seem to prefer to plunge from low-hanging limbs over pools. They are far from inconspicuous, with their blue and white plumage, belted breast, and topknot. Still, they do well on unwary salmon fry and fingerling trout, supplementing their diet with insects, amphibians, even small birds when necessary — a plentiful and varied fare that allows them to remain in the park year-round.

Other birds commonly seen along Olympic streams — mergansers, blue herons, wood ducks, spotted sandpipers, and killdeer — are habitués of the lower rivers, well below the fury and flash of rapids. Mergansers are found where river gradients flatten out and rapids become scarce. Much more common on the larger west-side rivers, notably the Hoh and Queets, mergansers feed on coho and Chinook salmon and steelhead smolt, as the young fish swim downriver toward the ocean. With their hooked, serrated bills and excellent diving ability, they are effective fishers and can reduce a fish population in a given stretch of river by half. As one fisheries biologist put it, "they're hell on coho." Mergansers nest in cavities in riparian snags as well as in banks and among boulders, brush, and roots. Their broods are large, up to 14 chicks, but 8 to 10 is more common. In winter, when salmon are scarce, they head out to salt water and feed on herring.

Wood ducks and great blue herons, while common in lower rivers and streams, are more often found in lakes, ponds, and tide flats. Wood ducks may well be the most striking nesting bird on the peninsula. The breeding male has a deep green and purple head with a downswept crest, striped and speckled throat, chestnut breast, and iridescent wings and back punctuated by an upswept tail. Even harlequins pale beside wood ducks, with which they are often confused. Wood ducks nest almost exclusively in riparian snags along slow streams, lakes, or ponds. They were hunted nearly to extinction in the late 1800s for their plumes, but populations have recovered since the turn of the century and now they are frequently seen on the peninsula. Wood ducks, harlequins, and other breeding birds may be the flashiest inhabitants of Olympic's riparian environments, but the distinction of most pervasive must go to an older and more secretive order, the amphibians.

The Double Lives of Amphibians

The first vertebrates to venture onto land, amphibians are the oldest land dwellers. Reptiles, birds, and mammals all came later. Amphibians still carry the signature of their watery origins; they keep a foot in each world (amphibian means "double life"). Most are hatched in ponds, lakes, and streams. In their legless larval stage, most amphibians live in water and breathe through gills like their ancestors the fish. Unlike reptiles and other true land dwellers, amphibians lack moisture barriers, either in their skin or their eggs, and so depend on moist environments to regulate body temperature. Adults of most species also return to water to breed. Few amphibians travel very far from water, and most venture out only at night. The wetlands, streams, lakes, and damp woods of the maritime Olympics provide ideal habitat for a variety of amphibians. And because nearly all of the park remains in a pristine natural condition, a number of sensitive amphibian species flourish here.

Fourteen species of salamanders and frogs are known to occur in the Olympics, including two species endemic to the peninsula, the Olympic torrent and Cope's giant salamanders. Isolated populations of Van Dyke's salamander and Cascades and tailed frogs are also found here; all are endemic to the Pacific Northwest. It's estimated that there may be up to 250 amphibians per acre in some areas of the Olympics; they may in fact be the major predators of invertebrates in many forest and stream communities. As is the case with many less-than-conspicuous species, scientists are just beginning to understand amphibians' importance in the forest ecosystem.

Fast-flowing mountain streams of the Olympics are home to both the endemic Olympic torrent salamander and the tailed frog. Torrent salamanders are small, about four inches, greenish brown above with distinctive yellow undersides. They blend well with the icy creeks and wet rocky stream margins they favor as habitat. Tailed frogs are also small, about the size of a treefrog, but lack the latter's dark mask (and exquisite singing voice). Tailed frogs thrive from sealevel creeks to mountain torrents, though higher-elevation populations are known to take longer to metamorphose into adults.

Both of these species are considered quite primitive by herpetologists. Stream-dwelling salamanders like the torrent salamander have been proposed as the ancestral form from which modern terrestrial salamanders evolved. The tailed frog is the most primitive frog in the world; its closest relatives are found in the rainy forests of New Zealand. Though plentiful in the park, both species are extremely susceptible to changes in their environment, and populations have been seriously reduced or eliminated in streams affected by logging.

Van Dyke's salamanders are rare and known from just a few locations besides the Olympics. Like the western red-backed and other woodland salamanders, Van Dyke's lays its eggs on land, usually beneath mossy stones and downed logs of riparian areas. The western red-back and the ensatina (a stouter salamander that lives in rotting logs and woody debris) favor divots and rodent burrows beneath the surface. In all three, the larval stage is completed within the egg, and the young emerge as miniature adults.

Rarest and certainly among the most secretive of Olympic's salamanders is the endemic Cope's giant salamander, known only to the Olympic Mountains and the coastal hills to the south. This species remained unknown until 1970, and only a few terrestrial adults have been observed since then. Most Cope's remain in streams and cold mountain lakes as larvae or in a subadult breeding form known as a neotene. Large enough to eat not only insects but fish and the eggs and larvae of other amphibians, Cope's are easily recognized. At maturity, adults reach eight inches and are strikingly marbled with hues of copper and slate.

Amphibians have become the focus of increasing attention over the past several years because of their emerging status as indicators of ecosystem health. Because their habitats are so highly specialized, amphibians appear to be especially sensitive to changes in their environment. Living in distinct microhabitats, they are generally unable to cope with rapid environmental changes by leaving the disturbed area. Some populations and even some species are small and isolated, yet they are important consumers of invertebrates — particularly in ecosystems that lack other kinds of insectivores.

Declines in amphibian populations worldwide have sparked concern among biologists, but disappearances from undisturbed and protected areas are particularly disturbing. Acid rain, increased levels of ultraviolet radiation, and global warming are among the possible causes for such losses. An obvious solution is to restore damaged natural systems such as streams, forests, and wetlands. For some aquatic habitats this can mean replanting riparian areas or mechanically placing logs and rootwads in streams to create spawning and rearing habitat. Olympic National Forest, the State of Washington, and volunteer groups working to save salmon have embarked on a series of watershed restoration efforts on peninsula streams that are already yielding positive results. On a larger scale, the National Park Service, along with other federal and state agencies and the Elwha S'Klallam Tribe, are pursuing a salmon recovery effort for one Olympic river system that promises to be the most successful ecosystem restoration in the Pacific Northwest.

Freeing the Elwha

The Elwha River salmon runs were legendary. In his book *Mountain in the Clouds*, Bruce Brown notes that as early as 1790 the Spanish explorer Manuel Quimper recorded purchasing a number of 100-pound salmon from nearby Indians. Reports of large Elwha Chinook remained common through the early years of this century as well, and it's estimated that at one time more than a quarter million pink salmon spawned in the Elwha River. The largest of Olympic National Park's watersheds, the Elwha was undoubtedly one of the most productive. The river supported 10 distinct runs of anadromous fish, including all five species of local Pacific salmon. It was one of the few rivers in the contiguous United States that harbored all the ocean-running fish species native to the Northwest.

The Elwha flows north out of the interior Olympics into the Strait of Juan de Fuca. In its lower reaches, where the river cuts through the resistant basalts of the Crescent Formation, it narrows into a series of steep canyons and shoulders through a notch called Goblin's Gate. Only the most vigorous wild salmon could make it through these canyons to spawn. Records suggest that Elwha Chinook may have remained at sea longer to gain the size and strength needed to surmount the canyons. They were certainly the largest Chinook in any Olympic river. An estimated 8,000 of them spawned in the Elwha each year. Equally impressive runs of coho, pink, sockeye, and chum salmon, steelhead, sea-run cutthroat trout, char, and bull trout spawned in its pristine mountain waters. For the Elwha S'Klallam people who lived at the river's mouth and for the wildlife species who inhabited the 175,000 acres of the Elwha watershed, these runs amounted to an incredible bounty as well as a dependable source of nourishment much of the year.

In 1913, all that changed. That was the year the Olympic Power Company began operations of its Elwha Dam.

Built to provide power for local industry, this was the first of two dams constructed on the lower river. In 1926, a second dam, the Glines Canyon Dam, was constructed several miles upstream. In spite of an 1890 statute requiring fish passage facilities on all dams, both dams presented a complete barrier to all salmon moving upstream to spawn. As early as 1911, while the Elwha Dam was still under construction, Clallam County game warden James Pike wrote the state fish commissioner. He reported "Thousands of Salmon at the foot of the Dam, where they are continually trying to get up the flume. I have watched them very close and I am satisfied now that they cannot get above the dam." Despite Pike's report and the protests of the S'Klallam people, whose very livelihood and culture was bound to the Elwha's salmon, state officials ignored the problem. Finally, the same year the dam went into operation, its owners reluctantly agreed to fund construction of a fish hatchery to mitigate the loss of native fish. But hatcheries were an unproven technology at that time, rife with problems. Within a decade the hatchery failed and was abandoned.

Of the 75 miles of available spawning habitat in the Elwha Valley, only the lower five miles are accessible to salmon today. And with the dams holding back the gravel needed to replenish spawning beds and warming impounded waters to dangerous temperatures in summer, even this short reach has become degraded. The river's sockeye salmon are now considered to be extinct, and chum and spring Chinook are rapidly approaching extinction. The once-great runs of pink salmon have dwindled to a few dozen and are now listed as runs of critical concern. Summer steelhead and sea-run cutthroat trout are also in steep decline. The sole beneficiary of the power generated by the dams today is the Daishowa America paper mill in Port Angeles; the dams supply about a third of the mill's energy needs.

The Elwha dams were products of an era of rapid industrial growth in the West, a time when few considered the environmental costs of their actions. With the subsequent declines of Pacific salmon throughout the region, and recent salmon listings under the Endangered Species Act, the true costs of such "free" sources of energy as hydropower have been realized. As always, it's the generations that follow those early boosters who must bear the cost of development. Beginning in 1968 the Elwha S'Klallam and other Indian tribes requested that the Federal Energy Regulatory Commission (FERC) address the problem of fish passage on the Elwha during its relicensing proceedings. The 50-year license for the Glines Canyon Dam expired in 1976; the Elwha Dam was never licensed. In the 1980s, as part of the relicensing process, the National Park Service and the U.S. Fish and Wildlife Service studied the feasibility of fitting both dams with fish passage facilities. The results showed little promise. Only five of the ten original fish stocks showed any chance of recovery with the retrofitting of the dams, and the prospects for even these were poor. By contrast, studies rated the chances of restoring nine of the ten runs as good to excellent if both dams were removed.

In 1986, a coalition of environmental groups intervened in the relicensing proceedings. The groups questioned FERC's jurisdiction to license dams in national parks and petitioned that the Glines Canyon Dam, which was located entirely in the park, be removed. In 1990, a consortium of federal and tribal agencies recommended that both dams be removed to fully restore the fishery. Two years later, with the support of a broad coalition of conservationists, sportsmen, public agencies, and tribes, Congress passed the Elwha River Ecosystem and Fisheries Restoration Act. To accomplish its goal of "full restoration" of the Elwha River ecosystem and native anadromous fish, the law authorized the Department of the Interior to acquire both dams and remove them if necessary to meet this objective. In 1995, the Department of the Interior and the National Park Service announced their intent to remove both dams, a bold decision that will return one of the Northwest's premier salmon-producing rivers to its natural state. Work is now underway that will right an 80-year wrong to a spectacular river system and the people who for thousands of years have made it their home.

The Elwha offers a rare opportunity to restore a major ecosystem to its unspoiled condition. Since the watershed lies almost entirely within

Olympic National Park, the habitat degradation from industrial logging, chemical pollution, and irrigation that have stymied salmon restoration efforts elsewhere in the Northwest are not an issue. I can't imagine another river system that holds more promise, and I can think of no finer investment than freeing this magnificent river system to simply do what it does best.

7
Life at the Edge of Land and Sea

For three days the winter storm lashed the coast. Dark skies leaned over mountainous waves, and gusts of wind whistled through the spruce limbs above my camp. From a perch well back among the trees I watched the waves crash against a rocky headland to the south. Once, whitewater exploded nearly a hundred feet upward, flinging spray into windswept headland trees, then drained off the cliff face in torrents. A half-breath later a second explosion burst like a cannon shot as the wave crashed through a tunnel in the headland. The two waves then chased each other as they fanned out across a protected cove, crested once more, and broke against the shore. It was seconds before the sound reached me —a muffled concussion I could feel in my chest.

For me, winter storms are the essence of this wild and rugged shoreline. The Olympic coast juts into the north Pacific like the prow of a ship, and the unbridled force of winter seas have sculpted it into a magnificent landscape. Storms give shape to the thick-limbed coastal spruces; they rake the tops of offshore islands, rearrange river mouths, and cut into coastal bluffs, sending huge trees crashing into the waves and leaving others leaning awkwardly like stunned giants. Like the glaciers and rivers inland, the erosive power of waves — and the drift logs, rock, sand, and trapped air they carry — batter and tear at the shoreline relentlessly. Storm-driven surf is a major force shaping both the character and ecology of the outer coast. Hammering at the edge of the continent, stripping away more than three feet of land each year in places, it quarries the coast into dramatic formations. The arches, tunnels, caves, and sea cliffs, as well as sea-stacks and rugged offshore islands that typify the Olympic coast, are the work of winter seas. And the intertidal communities that flourish within this powerful environment carry its signature.

The wildness wrought of the winter sea has probably helped protect the coast during the past century of development on the peninsula. Nowhere else in the contiguous United States has so generous a portion of wild coastline been preserved in its natural state, free of the roads, condominiums, resorts, and other trappings of commerce that pass these days for progress. The small Native American fishing villages at Neah Bay, La Push, Hoh, and Taholah remain largely unchanged. And the wild reaches of unbroken shoreline that stretch between them appear as they must have when the ancestors of today's coastal inhabitants guided their cedar canoes along the headlands, islands, and rocky coves.

The Shape of the Olympic Coast

The 62-mile coastal area of Olympic National Park stands apart. Not only does it remain undeveloped, its rugged nature lies in marked contrast to the low bluffs and long sandy beaches that typify the coastline to the south. There, thousands of tons of sediments washed into the sea by the Columbia River and carried by coastal currents give rise to extensive sandy beaches. But like the mountains a short distance inland, the shape and structure of the north Olympic coast reflects an uneasy marriage between the geologic forces that gave birth to the Olympic Peninsula and the climatic forces that conspire to worry it away.

The signature landforms of the Olympic coast — its islands, sea-stacks, and towering headlands — are formed of tough, erosion-resistant rocks. Like most of the core rocks of the interior peninsula, coastal rocks are sedimentary; sequences of sandstones, siltstones, and conglomerates (rounded fragments of rock in a matrix of sand or silt) were laid down

Janis Burger

on the ocean floor by ancestral rivers 20–40 million years ago. During uplift, as the ocean plate ground beneath the edge of the continental plate, these sediments were among the rocks scraped off the seafloor and plastered onto the continent's edge. Though bent, twisted, and broken into the usual lithic smorgasbord of Olympic rock, these coastal rocks (known to geologists as the Hoh assemblage) are not nearly as deformed as the interior core rocks that abut the submarine basalts of the Crescent Formation. Most of the headlands, sea-stacks, and islands that have held their own against the onslaught of the sea are composed of tough sandstones and conglomerates that were spared from severe folding and faulting during subduction and uplift.

Tough sandstone is slower to yield to the erosive power of the surf than adjoining melange rocks. This leads to the creation of the points, headlands, seastacks, islands, and the interspersed pocket coves and beaches that typify the north Olympic coast.

The more worked-over sedimentary rocks, where the tectonic stresses of subduction caused extensive folding and thrust faulting, resulted in a jumbled formation of unconsolidated rock appropriately called a melange. Here fragments and blocks of sediments and basalts are embedded in a soft mix of sheared and broken siltstones, sandstones, and clays. Melange rocks are much more prone to erosion. Yielding to the cutting power of the surf, they tend to form inlets and beaches rather than headlands and islands. The popular Second Beach and Third Beach near La Push as well as the long beach north of Hoh Head are carved into melange rocks. In contrast, James Island off La Push, Teahwhit Head between Second and Third beaches, Toleak Point, and Hoh Head are all composed of tougher sandstone conglomerate.

A notable exception to this story — indeed to the entire geologic history of the Olympics — can be found at Point of the Arches, a spectacular array of sea-stacks and arches on the northern coast. There, cliffs and seastacks composed of basalts, sandstones, and gabbro (a dark igneous rock formed at great depths) are about 144 million years old — twice the age of the oldest known rocks in the interior Olympics. It is the only exposure of its kind on the peninsula, and its origin remains a puzzle to geologists. Some theorize the rock arrived from the north by way of a massive submarine avalanche during the formation of the Olympics. Others suggest it was rafted here from the Klamath Mountains region in northern California as part of the crustal movement associated with the San Andreas Fault. That its origin remains a mystery is somehow in keeping with the haunting beauty of the northern coast.

The more recent 1.8-million-year history of the Pleistocene epoch is also written into the coastal text. During this time, when alpine glaciers were honing the peaks and carving the valleys of the interior and great ice sheets advanced repeatedly from the north, the coastline was also in flux. Outwash from the meltwaters of alpine glaciers covered most of the coastal area, a fact evidenced by the unconsolidated glacial drift seen along much of the coast today. About 125,000 years ago, sea level stood about 100 feet higher than today. Now covered by younger deposits of sand and gravel, the remnants of a flat, elevated, wave-cut terrace can easily be seen at the top of bedrock exposures along sea cliffs of the middle coast, as well as along the flattened tops of some offshore stacks and islands. Geologists have traced this Pleistocene terrace as far as a mile inland. It indicates the extent of an older coastline, much as today's wave-cut platforms surrounding headlands reflect present sea level. This same terrace cut into older glacial deposits as well. Deposits from more recent alpine glacier advances between 16,000 and 70,000 years ago lay atop the older terrace. This Ice Age story can easily be read along the coast south of the Quillayute River. North of there, the text has been revised by the more recent movements of the Cordilleran ice sheet bearing down on the peninsula from Canada.

During the maximum advance of the Cordilleran ice, around 14,000 years ago, sea level is estimated to have been as much as 400 feet lower than today, and a vast, ice-free plain stretched out along the coast. Dry westerly winds sweeping across the plain deposited between 3 and 15 feet of silt and sand along the coastal area. This fine, buff-colored layer is noticeable atop most coastal bluffs today. As the ice sheet retreated and the sea returned to its present level, wave action tore into the young glacial deposits with a vengeance. The islands and sea-stacks trailing off

As winter seas work away at cracks and weaknesses in coastal rocks, headlands are quarried into caves, arches, and eventually free-standing sea-stacks and islands. The view is south, through Hole-in-the-Wall, toward James Island just off La Push.

Janis Burger

from the coast today mark the past extent of the continent's domain, the flat tops of the larger islands the final, glacial-deposited remnants of the once-vast Pleistocene coastal plain.

Hikers along the coast sometimes notice wave-sculpted sea-stacks and arches reemerging from their casts of glacial drift like half-finished sculptures; Quateata Point, north of Second Beach, is one example of this process. At the same time, new caves, tunnels, and eventually arches are being carved from joints or weaknesses in coastal bedrock. All along the Olympic coast, offshore rock gardens like Quillayute Needles and Giants Graveyard, and larger islands like Tatoosh, Ozette, Alexander, and Destruction stand in silent testimony to that powerful dynamic by which the sea lays claim to the land. It is a process that pulses and ebbs through the ages like a tide, while in the relative stillness of the ocean's depths, riverborne grains of quartz, feldspar, mica, and clay deepen and compress into strata — laying foundations for mountains and coastlines yet to be born.

The Diversity of Coastal Environments

Although rocky shores dominate the Olympic coastline, the coast is remarkably varied. River mouths, pocket coves, sand and gravel beaches, and protected reefs in the lee of sea-stacks and islands create a complex array of intertidal environments, each teeming with life. The sheer diversity of coastal life forms may well surpass that of the inland ecosystem. And the marine shoreline of Olympic National Park is probably one of the most complex and diverse in all of the United States. This tremendous diversity is brought about by a number of factors. Some are products of the geologic forces that shaped the present coast; others find their origins in the wide and bountiful ocean from which the Peninsula itself was drawn.

The Pacific is an old ocean — much older than its sister the Atlantic — and its rocky shores have served as a refuge for marine plants and animals for hundreds of millennia. This relatively stable marine environment has allowed, in Thoreau's words, "a liberal amount of time" for evolution to perform its magic. Nutrient plumes from the Columbia River south of the Olympic coast and the Strait of Juan de Fuca to the north are freshened with runoff from Olympic rivers, bringing nitrates, phosphates, and other nutrients into coastal waters. Offshore, a complex pattern of currents and winds causes cold, nutrient-rich waters of the ocean depths to upwell onto the continental shelf and nearshore areas. This continuous surge of nutrients feeds free-floating microscopic plants in sun-warmed surface waters. They and the microscopic animals that feed on them (which include the larval stages of many larger animals) are called plankton. They form the base of the food chain in open-water ecosystems and also supply important food for filter feeders on the shore. Twice each day the sea floods the lush intertidal areas of the North Pacific with this plankton-rich broth.

The Olympic coast's biological diversity is also enhanced by climate. Unlike many rocky shorelines elsewhere in the temperate zone, the Pacific Northwest coast enjoys a consistently mild maritime climate. Temperatures do not vary greatly from season to season. The same coastal fogs that nurture the Olympic rain forest in summer cloak the shoreline and temper the desiccating effects of the July sun. Hard freezes, which can wreak havoc on intertidal organisms exposed during low tides, are rare; sea ice, which can seriously abrade intertidal communities, is nonexistent. The Olympic coast's location midway between California and Alaska allows it to harbor species common to both boreal and temperate marine environments. All this combines with the variety of habitats and the relative absence of human-caused disturbance to create a realm of exceptional productivity and beauty.

In summer, low tides and calm seas grant us terrestrial types passage among the tidal pools, surge channels, and seaweed-covered rocks to glimpse communities that stay hidden for much of the year. Researchers have estimated that there are at least 130 plant and 180 animal species inhabiting the narrow zone between Olympic tides, and these figures do not take into account the many birds, sea mammals, and land animals that also use the Olympic coast. This high level of diversity is reflected throughout the coast's environments, rocky shores, cobble benches, and sandy beaches, but along exposed rocky shores it reaches its fullest expression. Rocky shores hold the greatest number and variety of marine organisms — from cormorants and turnstones to anemones and mussels — of any coastal environment. By observing the ways key intertidal plants and animals interact and respond to competition, predation, and disturbance, we can gain a better understanding of this dynamic environment and a keener appreciation of its place within the larger Olympic ecosystem.

A Ramble Through the Rocky Intertidal

If winter storms punctuate one end of the coastal year, summer explorations of the low tide zone illuminate the other. I like to start early on summer mornings, well before the first rays of sunlight stir the fog from the tops of coastal spruces. The cries of western and glaucous-winged gulls announce the turn of the tide as I clamber around the high rocks of the spray zone to see what I might scare up. Like the the plants and animals of the interior mountains, marine organisms of the rocky coast arrange themselves in layers or zones in response to environmental conditions, including the biological conditions of competition and predation. On rocky shores, these zones are compressed into tens rather than thousands of vertical feet, all reachable in the lull of an outgoing tide.

The Spray Zone

The spray zone is moistened only by the highest tides and rain, and last night's quiet high didn't bring much action. Temperature and salinity vary greatly, and the few organisms that live here have learned to cope with extremes. A band of dark lichen grows along the upper end of the spray zone, and bright green sea hair algae mark the freshwater seeps. Below them, clusters of common acorn barnacles *(Balanus glandula)* crowd shoulder to shoulder with small acorn barnacles. Like many species, common acorns are found throughout the intertidal area. They would crowd the small acorns out completely if not for heavy surf and wave-tossed driftwood that clear patches of them, allowing small acorns to recruit. Barnacles, lichens, and other dwellers of the high shore can thrive higher up on the wave-exposed sides of rocks; they are found slightly lower on lee sides. An occasional rock louse darts across a rock; this small marine isopod shows a decided leaning toward dry land. I notice a few checkered and Sitka periwinkles huddled in a wet crevice along with some tiny finger limpets. Limpets are tough conical mollusks that clamp themselves to rocks with a muscular foot. They can stick with a force equal to 80 pounds — a handy trait when winter seas pound the coast. By sealing themselves into their "home scars," which they wear into the rocks, some limpets can also stay wet during low tides. Farther down among the boulders, larger speckled limpets feed on algae growing on the undersides of rocks.

The High Tide Zone

As I climb down into the high tide zone, thick beds of common acorn barnacles cover the rocks — the small acorns of the splash zone are less apparent here. Larger limpets also appear; brown and white shield

Among the most common seaweeds of the upper and middle intertidal zones, thick beds of rockweed provide a wet protective cover for limpets, snails, and other intertidal species during low tide.

Janis Burger

limpets are common, and flatter plate limpets sport tufts of algae that allow them to blend inconspicuously with their surroundings. Limpets make their living by scraping algae from the rocks with toothed feeding organs called radulas. Farther down in this zone, whelks use radulas to prey on acorn barnacles and mussels by boring holes through their shells. In fact the presence of predatory whelks inhibits the downward spread of acorn barnacles. Where the acorns become thin, I notice the dull gray husks of larger thatched barnacles, and dense clusters of black turban snails congregate among the cracks and crannies in areas protected from the waves. Turban snails can live for 40 years, as measured by growth rings. When they die, their shells are commonly appropriated by hermit crabs. Lacking a shell of their own over their soft, vulnerable abdomen, hermit crabs must trade old shells for new frequently as they grow. They seem continually preoccupied with their search for new shells, and, as in certain desirable neighborhoods elsewhere, contests over available homes can become quite pitched. Hairy and glandular hermits inhabit the wetter parts of the high tide zone; both protect their newly acquired real estate with stout oversized claws.

Like their alpine counterparts, organisms of the high tide zone must be able to cope with fairly demanding conditions — not only heavy wave action but long periods of exposure to air, drying winds, and sunlight. Some, like the rockweeds, have become specialists. Thick beds of brown rockweed drape the substrate of the middle high tide zone, and the pale air bladders at the tips of their fronds pop wetly under my feet. When damp, rockweeds photosynthesize better in air than in seawater — one reason I find them this high on the shore. And like oaks and a few other terrestrial plants, rockweeds contain tannins, which may inhibit grazing by periwinkles. Moving aside the fronds I find congregations of limpets, snails, and isopods living beneath rockweeds, eluding predators as well as escaping drying heat. Thin, nearly transparent fronds of bright green sea lettuce wave lightly in the nearby current, and low-growing tufts of sea moss, looking like forest moss, wedge among the rocks and cobbles. Purple shore crabs scuttle sideways from rock to rock. If I inadvertently corner one of these tiny creatures away from cover, I'm always a bit startled to see the fierceness with which it bares its small pincers. Inevitably, I'm the one who retreats.

The Middle Intertidal

By early morning, the lush seaweed gardens of the middle intertidal zone begin to emerge, and rivulets of seawater pulse back and forth at their fringe. This zone is bathed in seawater about half the time, and seaweeds here grow profusely. As algae, which lack the roots, leaves, flowers, and stems of herbaceous land plants, seaweeds are classed low on the evolutionary ladder. Their "roots" are really holdfasts that keep them attached to the bottom, their "leaves" blades that absorb minerals from seawater as well as photosynthesize sunlight. But in spite of their simple structure, these ocean plants embody a variety of shapes and colors, and they vary widely in reproductive strategies as well as habitats.

I find fewer rockweeds in the middle intertidal, but bright bundles of sea lettuce linger, sometimes in thick beds. Sea lettuce here is joined by other garden-variety seaweeds — sea cauliflower and sea cabbage among them. The cauliflower resembles a brownish version of its garden namesake, while sea cabbage spreads long split leaves over the rocks. In the calmer waters of Puget Sound

Common organisms of the intertidal zone.

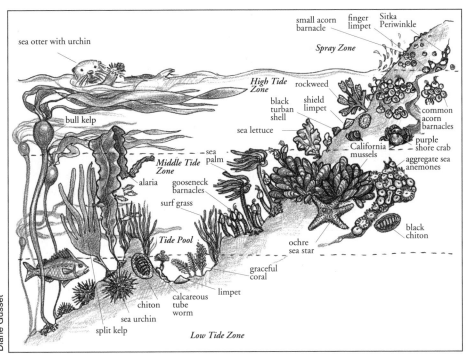

Diane Gusset

it more closely resembles a cabbage. Sea sacs lift two- to three-inch clusters of water-filled saclike tubes into the current, and black pine, which is actually brownish in the water, trails bristly needlelike branchlets in the outgoing tide.

On the steeper, more exposed, wave-swept shores, a true indicator plant of the open coast, the sea palm *(Postelsia palmaeformis)* occupies cleared patches in mussel beds. Its presence is a warning for hikers to proceed with caution. With flexible stalks up to two feet tall topped by long shaggy blades, sea palms droop when out of the water, resembling miniature palm trees. Their tough holdfasts grow over barnacles and mussels, providing a firm anchor in battering surf. Sea palms are annuals, and fall storms strip them away, along with the plants and animals caught in their holdfasts, creating new attachment sites for their spores.

Such wave-cleared patches enhance community diversity in mussel beds by allowing new plants and animals to get a start. If patches are small, limpets and chitons may keep them free of algae, and barnacles can set. If patches are larger, islands of algae will form in their centers, and sea sac and other seaweeds soon follow. The plant communities of the middle intertidal harbor a greater number of animal species than are found in the high tide zone. But the most characteristic organism of this zone, the California mussel, creates its own microenvironment.

Mussels and Sea Stars

If there is a signature species of the coast's rocky intertidal zone, it's the California mussel *(Mytilus californianus)*. As the tide recedes about halfway out, I find wide bands of these large bivalves cobbling the rocks; they seem to crowd out everything but scattered clumps of gooseneck barnacles. The mussels range from dark blue to brown and black in color, and the shells of the larger ones are often encrusted with acorn barnacles and limpets. These mussels can live up to 20 years. Quiet hissing and popping sounds permeate the bed as water drains away, and I try to step lightly over this crowded neighborhood. Though they appear a monoculture at first glance, these beds can be several layers deep, and they offer a rich protected microclimate for as many as 300 other species. Tube and ribbon worms, small clams, snails, limpets, and crabs all find suitable habitat in the interstices of mussel beds and feed on the detritus, algae, and small animals that collect there.

Mussels attach themselves to rocks with tough elastic byssus threads that can be absorbed or respun, affording them a small degree of mobility. I notice a few smaller, smooth-shelled, edible blue mussels scattered among the beds, but they generally prefer more protected sites in the lee of sea-stacks and islands or above California mussels on shore. Blue mussels are much more mobile than California mussels and are quick to colonize wave-cleared patches. Eventually, the larger Californians outcompete them on the open coast; they overgrow the smaller mussels and crush them out. Blues find more favorable habitat in calm bays and the inland waters of the Strait of Juan de Fuca and Puget Sound.

The other contenders that wedge their way into mussel beds are gooseneck barnacles. The larvae of both goosenecks and California mussels prefer to settle among the byssus threads and stalks of their own kind. Like their thick-shelled neighbors, goosenecks are well adapted to life on the stormy coast. Their short tough stalks, tipped with pearly light gray segmented plates, give and bend with the surf. Unlike other barnacles that pull in their food with featherlike cirri, goosenecks simply strain runoff from wave-inundated rocks, making them well suited to the vertical walls and undersides of rocks the mussels force them to occupy. Although gulls are known to feed on them, the major predator of both gooseneck barnacles and California mussels is the common ochre sea star *(Pisaster ochraceus)*.

Looking near the water line just below the mussel beds, it's not hard for me to find them. Purple or ochre sea stars are the most common sea stars on the open coast. Stout, five-rayed starfish, they can grow to more than a foot across. As its name implies, *Pisaster* can range from ochre-orange to bright purple with shades of tan and brown in between. Ochre is the color I see most often on the open coast; purple is more common in Puget Sound. Their bright colors make them stand out at low tide, and I soon notice several more clumped beneath an overhanging ledge by a pool. Though they feed on barnacles, limpets, and snails, the prey most preferred by ochre sea stars are mussels.

An exquisitely evolved predator, sea stars cover mussels and pull their shells open with powerful arms and strong tube feet. Once a mussel is open even a crack, the sea star can insert its stomach into the shell and digest it. Sea stars are equipped with light- and chemical-sensitive tentacles at the tips of their arms and have rough backs, rasplike to the touch. Their backs are covered with microscopic pincers that keep other organisms from settling on them. *Pisaster*'s unusual feeding habits limit their feeding to high tides, and they show a strong aversion to air and sunlight. This keeps them generally low in the intertidal zone — and keeps mussels growing higher on the rocks generally safe from sea star predation. In fact, sea stars define the bottom edge of California mussel beds. When they are removed, mussels rapidly extend their ranges deeper into the intertidal. By limiting the sizes of mussel beds and clearing patches of bare rock within them by predation, sea stars play a pivotal role in the ecology of the middle intertidal zone. Some 25 other species — including a number of barnacles and algae — depend upon ochre sea stars to create open patches in mussel beds.

Predation by purple or ochre sea stars creates openings in beds of California mussels and enhances intertidal diversity. Some 25 intertidal species benefit from the predation of sea stars on mussels.

Another interesting intertidal relationship is that between the sea stars and a seaweed, sea moss (*Endocladia muricata*). Among the first algae to colonize the clearings that sea stars carve in mussel beds, sea moss provides an

Janis Burger

ideal base for mussel larvae to settle. Limpets are the primary grazers of sea moss, and so limit attachment sites for mussels. Conveniently, ochre sea stars also prey on limpets, and by trimming their numbers, help insure a future supply of their preferred prey, California mussels.

Pisaster is a quintessential "keystone species." These are organisms, usually high on the food chain, that through their activities exert an influence on the composition of their communities disproportionate to their numbers. Scientists studying the rocky intertidal have found that the lower ranges of a number of species are set by predation as well. The distribution of acorn barnacles is limited by predatory whelks, and the lower limits of some seaweeds are probably set by the grazing of black chitons. In contrast, the upper limits of many species' ranges are most often determined by environmental factors — exposure to waves, air, and desiccating wind, and lethal warming by sunlight.

Poking around, I notice a few six-rayed sea stars occupying the same niche as ochre sea stars. They have the same feeding patterns as ochres but are much smaller (three to four inches), and feed only on smaller individuals — barnacles, limpets, snails. They exert much less influence on the ecology of this zone. As one biologist told me, *Pisaster* drives the middle intertidal; remove it and diversity crashes. Unfortunately, illegal collecting of starfish in some heavily visited intertidal areas outside the park, such as Pillar Point on the Strait of Juan de Fuca, has resulted in just that.

Along the wet edges of mussel-cobbled rocks, I find clusters of aggregate sea anemones hunched together like Easter eggs in a carton. Their small pale green and pink "flowers" have closed as the tide has retreated, and they are covered with bits of shell and gravel. Able to withstand the pressures of wave action, sunlight, and wind, these tough, two-inch-high animals will wait for the returning tide to deliver zooplankton and other prey into their sticky tentacles. Though they frequently live next to mussel beds, most anemones aren't preyed upon by starfish. In fact, some benefit from starfish predation by picking up table scraps from the starfishes' meals. Aggregate anemones reproduce asexually by cloning; larger individuals divide as often as once a year. Interestingly, one cloned colony will keep another at "arm's length" with the stinging capsules at the tips of their tentacles.

I often find black chitons in association with anemone colonies. A tough mantle covering most of their shell allows these chitons to remain exposed to air and sunlight for long periods. Black chitons are easily identified by the eight overlapping plates along the spine of their shells as well as their medium size (three inches). A tough leathery foot keeps chitons fastened to rocks and allows them to move about and feed on diatoms (microscopic one-celled plants) and larger algae. They're also quite fond of sea cabbage. By clearing this fleshy seaweed from the rocks, chitons make room for coralline algae to grow. These are the lovely pink plants, both branching and crustlike, that lend tide pools an almost otherworldly air.

The Low Tide Zone

By midmorning the tide has drawn well back from the shore, revealing a broad low rocky platform glistening with dark seaweeds. Shallow ravines or surge channels bisect the rocks, and tide pools abound. These rocky depressions where seawater lingers at low tide are found throughout the intertidal area, but the tide pools of the low tide zone hold a special magic. The shapes and colors of low-tide dwellers, particularly those revealed during the "minus" tides of spring and summer when water recedes more than 10 feet below mean high tide, are most vivid. They suggest an even richer abundance and diversity of species that remains hidden in the subtidal zone. The sound of a subdued surf lapping near-by rocks and the brevity of each visit — dictated by the returning tide — heighten the sense of discovery.

Peering into a Tide Pool

By allowing marine creatures to stay active during low tides, tide pools act as lenses through which to glimpse the dynamics of intertidal life. Barnacles and tube worms wave feathery feeding structures, and anemones open their flowerlike tentacles to the flow. Sheltered from the brunt of wave action, less resilient plants like surfgrass and coralline algae flourish in these natural aquaria, and interactions between organisms can be easily observed. Limpets and chitons can be seen slowly scraping the rocks for food; a shrimp may nibble a bit of dead fish; a whelk may drill away at a mussel. My first pool yields what many consider to be the most strikingly beautiful tide pool inhabitant, the red and green anemone. Algae living inside their tissues give most anemones their color, but the red streaks of this anemone's column and the pink to reddish stripes of its tentacles set it apart. The relationship between algae and their anemone hosts is symbiotic; anemones use carbohydrates fixed by the algae, and algae use carbon dioxide from anemones.

A relative, the giant green anemone, is similar in size (up to 10 inches in diameter), but tends to favor more wave-swept environs. I often find them in surge channels where food is carried to them by tidal flow, as well as alongside mussel beds where, like aggregate anemones, they pick up meals loosed by starfish. Like all anemones, the giant green's tentacles close over its prey — crab, mussel, or limpet — and guide the prey into its gullet; then the anemone "swallows" it. Later the anemone expels undigested bits of shell. Green anemones can become quite old — some are thought to reach ages of 1,000 years. In areas where waves are quieter, they are smaller and markedly fewer in number.

A tiny, almost iridescent anemone, the brooding anemone, also brightens some tide pools. They are only one-half to two inches in diam-

Large numbers of sea urchins graze on kelp beds and other seaweeds in tide pools or offshore waters. The red sea urchin is the largest on the Olympic coast and prefers the shelter of tide pools and protected waters. Sea urchins are a favorite food of sea otters.

Ruth Kirk

eter. I find them hidden among blades of surfgrass where, like the six-rayed sea star, they brood their young until the offspring are large enough to survive on their own.

Spiny sea urchins add bright splashes of purple and green to some of the larger tide pools. The red sea urchin of the outer coast is the largest found here; it can grow to more than five inches across, and its spines can jut out another three inches. Despite their formidable appearance, sea urchins feed primarily on algae. They also use their sharp teeth and spines to scour indentations in soft rock where they are protected from tides and currents. Where wave action is severe, smaller purple sea urchins carve deep pockets.

Sea urchins are known to feed heavily on kelp throughout much of the low tide and subtidal zones. In southern California, an overabundance of sea urchins in past decades (brought about by the extirpation of their major predators, the sea otters) led to the destruction of large offshore beds of bull kelp. When the urchins were artificially controlled, kelp beds recovered. For years, sea urchin gonads have fetched very high prices in Japan, where they are considered a delicacy. In areas where urchins have been commercially harvested, kelps tend to dominate seaweed communities, and sea urchins may take years to reestablish themselves. But urchins are prolific, even among marine animals; a female can produce 20 million eggs in a single season.

Crouching low and peering beneath tide pool ledges, I find the long white casings of calcareous tube worms twined over a crusty bed of coralline algae. I don't see their feathery fans out sweeping the water for plankton, but nearby, clusters of smaller feather duster worms open thick multicolored mops that look like — well, like miniature feather dusters. Multiple layers and tiers of life deepen tide pools with vivid colors and textures. Coralline algae range from bright pink to pale purple in delicately branched or platelike form. The segmented calcified tissue of branched corallines lets them sway gently in the current. More than 10 species of coralline algae are found in Olympic tide pools. Graceful coral seaweed spreads pink lacelike branches, while coral leaf seaweed is almost skeletal in comparison. Many of these delicate algae depend on chitons and other grazers to prune back fleshy seaweeds that would otherwise choke them out. They, in turn, feed communities of specialized grazers: lined chitons eat encrusting corallines; black chitons feed on branched.

Multihued rock crust seaweeds, brilliant sponges, and chartreuse hydrocorals carpet tide pools with dazzling displays of color — much of it sheer biological extravagance. Unlike the coloration of most terrestrial animals, that of many marine organisms seems to serve no adaptive purpose. An exception to this may be the small nudibranchs that move among the seaweeds. Their extravagant colors may advertise bitter tastes or even toxicity. One, the opalescent nudibranch, is particularly stunning. With cerata tipped with burnished scarlet and electric blue stripes lining its sides, it could have just wriggled out of Disney's *Fantasia*. Other nudibranchs blend inconspicuously with their surroundings, much like the small quick-moving shrimp or darting tide pool sculpins that shoot under rocks as I approach their pools and stay hidden no matter how hard I try to search them out.

Tangled Mats of Seaweed

Beyond the tide pools, the rocky reefs of the low tide zone stretch out to the low waves in a tangled mat of seaweeds. Stalks and blades eddy and bob as wavelets break against the rocks. The plant associations of the low tide zone are commonly arranged in multistoried communities, not unlike the layers of an old-growth forest. Large kelps or sometimes surfgrass form the upper canopy. Kelps are tough brown seaweeds often growing in dense colonies just offshore. Surfgrass is one of the few vascular plants of the intertidal, getting its nutrients through roots and reproducing by flower and seed like inland species (seagrasses may have found their way back to the ocean by way of rivers and tidal estuaries). An understory of corallines and various red algae thrives in the shelter of the kelps, and a layer of red algae often encrusts the rock surface below. Many of the more than 100 species of algae that grow in Olympic's intertidal areas appear to require the cover of larger seaweeds for protection from wave shock as well as exposure during low tides.

Grazing by herbivores such as mollusks does not seem to significantly hamper plant growth on the Olympic coast. Limpets, which are among the most numerous grazers elsewhere, are abundant only in the middle and high tide zones here, and most chitons are generally too small and slow moving to have a significant impact on intertidal plants. The large

Janis Burger

Like virtually all seaweeds growing off the coast, the common kelp, feather boa, is an alga. Algae absorb nutrients from seawater and photosynthesize sunlight through their blades. Only a few marine plants, such as surfgrass and eelgrass, have roots, stems, and flowers.

leather chitons, which scrape portions of the rocks bare in parts of Puget Sound, are less abundant on the outer coast; other grazing species may be fewer in number as well.

This relative paucity of herbivores may be partly explained by predation. Gulls, oystercatchers, and starfish are all known to feed on these species; oystercatchers in particular have been known to put serious dents in limpet populations within their territories. By keeping herbivorous species in check, the predators on the Olympic coast may be partly responsible for the lushness of these low-tide gardens. Of course, a healthy flow of nutrient-rich water and the early-morning timing of low tides during the warmest months (covering seaweed beds with seawater during the heat of the day) are also important factors.

Among the algae most often found in the lower reaches of the low tide zone are several species of kelps. Split kelp is fairly common, its blades cut into ribbonlike strips. Alaria grows up to 10 feet long and has rich brown blades with pronounced central ribs and furled edges. Ornate, fringed feather boas live up to their name, and wide blades of yellow-green sugar kelp drape the wet rocks. Just offshore, bull kelp and perennial kelp grow in dense "forests" that mirror the upland rain forest. These kelps can grow to more than forty feet long. Their dense, tubelike stems, or "stipes," and flowing blades provide shelter and offer a range of complex habitats for a variety of fish, as well as sea otters and other marine organisms. Bull kelp favors somewhat protected areas, while perennial kelp does well in the most exposed rocky reaches. Thriving year-round on the open coast, perennial kelp serves as critical habitat for sea otters, who feed among its stalks, raft together in its beds, and "tie into" it during rough weather. When cast ashore as wrack, both kelps form huge, tangled drifts on the beach. Actually, most kelps are more easily seen when washed up after fall or winter storms. Their littered remains, strewn along the high tide line, harbor busy colonies of beach hoppers and other tiny scavengers. They, in turn, are an important food for a variety of shorebirds. Like downed logs in the rain forest, rotting kelp plays an important role in cycling nutrients back into the nearshore system.

One last seaweed, the rainbow seaweed, deserves mention — not only for its abundance along rocky shores but for its sheer beauty. Its smooth wide blades are coated with a lustrous cuticle which imparts to this red alga a shimmering blue-green iridescence. Its rainbow colors shift with the angle of the viewer or as sunlight moves across the sky.

On the lowest of tides, kelp fronds drift on slack offshore waters, and creatures more common to the subtidal area sometimes make themselves known. During other explorations at very low tides I've been lucky enough to spot bright orange sunflower stars. Two feet or more in size and with as many as 24 rays, sunflower stars are fleetest of foot among the sea stars, and they inspire in their prey dramatic acts of escape. Scallops will clap their valves together frantically to swim away; large sea cucumbers have been known to "gallop," and common cockles will kick their foot in a rapid retreat across the substrate.

The smaller sun stars are a striking orange to rose color with gray or blue stripes extending from the center of their backs down the lengths of their 10 rays. I found neither of those on this visit, but the much smaller blood stars were out and about, their five slender rays a brilliant orange-red. Like brooding anemones, these sea stars share the unusual habit of brooding their young, keeping them nearby until they are large enough to fend for themselves.

I keep an eye over my shoulder, watching the two or three high rocks

I used to skip across channels, making sure they're still there for my retreat. Like darkness on a forest trail, the change in tide can come suddenly, often when I've got my face in a tide pool and my hind end to a wave. As the tide turns and I begin to retrace my route to shore, I find a gum boot chiton — looking quite like the castoff sole of an old rubber boot — clamped to the wall of a channel. Much larger and more leathery than the hairy and mossy chitons also found here, the gum boot is the largest chiton in the world. I hear the clicking of crabs hidden beneath seaweeds as I hop and skip back to shore, kelp crabs and other spider crabs no doubt, stealthily working their weedy worlds. Spider crabs like the decorator crab adorn themselves with glued-on bits of seaweed and shell. Hairy and helmet crabs sprout less elaborate growths of seaweed. No doubt all are grateful for the return of the tide and the departure of unwelcome visitors like me — turning over rocks, poking through seaweeds and making a general nuisance of myself. It's probably for the best that these lowest tides come only every month or so in summer.

Cobble, Driftwood, and Sand

Toward the southern end of the Olympic coast, south of the Hoh River, sand and cobble beaches become more common. The forests that thatch the roofs of sea-stacks and headlands to the north step down gently here, and spruce limbs bow to the tide line. Beneath them, a band of silvery driftwood marks the fluid and shifting edge between land and sea. At a few places along the coast, such as Rialto Beach, this dynamic edge is quite vivid as the forest retreats from the restless advance of the surf. Just past the parking area at Rialto, dead and dying spruce trees rise from a litter of wave-tossed cobbles. Mossy logs, tree roots, and brush of the forest floor poke out from beneath heaps of stones cast up by winter storms.

Cobble or shingle beaches like these are among the most actively shifting on this restless coast, and they provide the poorest habitats for organisms. Constantly shuffled and ground together by wave action, cobble on the open coast supports little of the marine life found on cobble beaches in the more protected waters of the strait and Puget Sound. Sandy beaches provide much more hospitable homes.

The smooth sandy beaches at Kalaloch, Ruby Beach, Second Beach, Sand Point, and Shi Shi are among the most beautiful stretches of the Olympic coast. Long breakers curl toward the shore, and gulls wheel above the blowing spray. Like exposed cobble beaches, the sandy beaches that dimple pocket coves and line unbroken stretches of shoreline seem rather lifeless at first. But feeding gulls and flocks of western sandpipers, sanderlings, and plovers that skirt the water's edge hint at the life hidden beneath the tide-washed sand.

Like the rocky headlands and cobble beaches that separate them, sandy shores receive the full force of stormy seas. The jumble of huge driftwood logs lining their upper reaches attests to the power of winter surf. Though the logs help stabilize the upper reaches, most sandy beaches are pummeled, churned, and abraded relentlessly by the pounding surf. Even adjoining rocky areas are scoured by the action of waves laden with sand and pebbles. The only large organisms that can survive in this tumultuous environment are those that can burrow down into the sandy substrate — such as crustaceans and worms — or those able to move quickly up and down the beach with the tides.

Most noticeable among the latter are the ubiquitous beach hoppers. Beach hoppers, or sand fleas as they're sometimes called, burrow into shallow sand or hide beneath tangles of seaweed and pieces of driftwood to keep moist by day. At night they turn sandy beaches into moonlit sock hops, as they pop about scavenging for food. These small creatures actually rely on the moon to navigate back to their burrows on the upper beaches.

Opossum shrimps and waterline isopods drift in with the tides to feed in the shallows; they in turn serve as food for larger predators. Snails leave wriggly trails across the wet sand as they scavenge for food. One of the most striking is the purple olive snail. Less than an inch long, its smooth, shiny, purple-gray shell may be the loveliest on the coast. Another favorite among beachcombers is the sand dollar. The small flat hollow disk cast up on shore is the skeleton of this urchinlike animal. It plows the sandy offshore bottom, picking out bits of organic matter. The five-petal pattern inscribed on the sand dollar's shell resembles a sea star, a hint that these species are closely related.

Sandy intertidal areas provide no holds for seaweeds, and large seaweeds cannot grow here, but the incoming tide brings microscopic phytoplankton, which feeds many sand dwellers. The same tides also bring predators such as flatfish, crabs, even starfish to feed on the submerged life. Razor clams inhabit the lower end of the intertidal, coping with their turbulent environment by rapidly burying themselves. Thrusting downward with a powerful foot and expanding its tip, razors contract their foot muscles to pull themselves deeper into the sand. As anyone who has dug razor clams can attest, these long slender bivalves can be amazingly fast. Razors, which are prized by clam diggers, are found mostly along the broad sandy beaches of the Kalaloch area, at the lowest of low tides.

Razor clam harvests in the park are managed by the Washington Department of Fish and Wildlife, but as the number of visitors to the coast increases, managers are reassessing long-standing recreational harvests of clams and a variety of other intertidal organisms. Some species are diminished or disappearing in heavily used areas like Kalaloch, and researchers and park managers are concerned that even recreational activities in the intertidal zone could pose a threat to diversity. Sport fishermen harvesting tube worms for bait have already put a serious dent in populations in the Kalaloch area. Illegal collecting of starfish and other organisms could alter intertidal communities and affect other species as well. Overgathering of mussels could hurt the many species that find habitats within mussel beds.

Currently, park rules prohibit collecting tube worms, gooseneck barnacles, starfish, and other creatures. But some scientists and conservationists feel that any harvesting of intertidal species, including clams, mussels, and nearshore fish, should be prohibited within Olympic National Park — just as hunting, cutting trees, and gathering wildflowers are prohibited in the park's interior.

The Olympic coast is a realm of transformation, an ecotone between a unique terrestrial ecosystem and the larger North Pacific world of which it is a part. That a relatively intact forest ecosystem should be complemented by a protected marine area with the richness of the Olympic coast is the rarest of good fortunes. We owe it to future generations, as well as to the life forms themselves — the numerous birds and mammals that share the coast and the myriad organisms that inhabit its intertidal edge — to keep this wonderfully complete ecosystem intact.

Lichens color a basalt outcrop on Blue Mountain. The jagged spires of the Needles, in the distance, are part of the Olympics' inner basalt ring, a section of submarine lava sheared off by faulting and folded in among mixed sedimentary rocks. *(Keith Lazelle)*

Lake Constance and Avalanche Canyon below Mount Constance in the east Olympics are carved into thick marine basalts of the Crescent Formation. *(Keith Lazelle)*

Left: As winter snow melts back from High Divide in the central Olympics, fields of avalanche lilies flutter in the summer breeze. The glaciers of Mount Olympus in the background are still laden with winter snow. *(Keith Lazelle)*

Above: As soon as snow melts from the high mountains, even rock faces and ridge tops burst into bloom. Spotted saxifrage, arnica, and Davidson's penstemon grow among splintered shale near Gray Wolf Pass. *(Keith Lazelle)*

Right top: Dawn light on Mount Olympus. Collecting more than 100 feet of snowfall each year, Mount Olympus supports the lowest-elevation system of glaciers in the 48 contiguous states. *(Keith Lazelle)*

Right bottom: The crevassed mouth or "snout" of Blue Glacier on Mount Olympus. Like many large glaciers in the Olympics, Blue Glacier has been in fairly steady retreat since the 1920s. *(Keith Lazelle)*

Above: Plants of the alpine and subalpine zones have adapted to cope with quickly changing conditions. An early summer snowfall melts back from the blossoms of spreading phlox. *(Keith Lazelle)*

Left top: Lake Crescent after a rain. As winter clouds rise over the mountains, they bury the high country in snow, but in the lowlands precipitation falls mainly as rain. Lake Crescent, 17 miles west of Port Angeles, receives more than 80 inches of rain a year. *(Keith Lazelle)*

Left bottom: Sunset from Blue Mountain. The rounded foothills west of Blue Mountain reflect shaping by the Juan de Fuca lobe of the Vashon ice sheet 15,000 years ago. One lobe of ice flowed west over these low ridges and foothills toward the Pacific, carving the Strait of Juan de Fuca; another pushed south along the east Olympics to carve Puget Sound. *(Keith Lazelle)*

Winter on Gray Wolf Ridge. In the heart of the northeast rain shadow, stunted lodgepole pines reach toward treeline on dry south-facing slopes. An adaptable tree, at home
in stressful environments, lodgepoles also thrive in a few locations along the Pacific
coast and Puget Sound. *(Keith Lazelle)*

Right top: One of the most beautiful of Olympic's endemic plants, Piper's bellflower
grows in cracks and crannies in the rocks of the high mountains. Biologists suspect it is
a "paleoendemic," a plant whose former range was erased by the Cordilleran ice sheet.
Its closest relative grows more than 600 miles to the north. *(Keith Lazelle)*

Right bottom: Another Olympic endemic, Flett's violet also prefers rocky crevices.
Blossoming in early summer, it brings a special delight to alpine scramblers.
(Keith Lazelle)

Tufted saxifrage, mountain oxytropis, and spreading phlox are typical of the hardy low-growing plants found in Olympic's alpine zone. Summers are short and conditions often severe in this realm above the trees. *(Keith Lazelle)*

Moss, licorice fern, and the hanging clubmoss selaginella drape the limbs of a big-leaf maple in the Queets rain forest. Epiphytic growth on some rain-forest trees can amount to as much as four times the weight of a tree's foliage. *(Keith Lazelle)*

Dense beds of Oregon oxalis carpet lowland forest floors in the western Olympics but virtually disappear in dryer east-side forests. The small inconspicuous flowers of oxalis bloom from early spring to late summer. *(Keith Lazelle)*

Right top: The sight of Roosevelt elk in valley forests always brings a thrill. It was to protect elk habitat that Theodore Roosevelt created Mount Olympus National Monument in 1909. Currently, an estimated population of 5,000 elk inhabit the park. *(Keith Lazelle)*

Right bottom: The canopy lichen lobaria serves an important role in the forest ecosystem by absorbing valuable nitrogen from the atmosphere and making it available to the forest. When it falls to the forest floor, it also serves as an important seasonal food for elk. *(Janis Burger)*

Above: Its slightly drooping limbs and short flat needles make western hemlock easy to recognize. Hemlock's tolerance to shade makes it the major understory tree in Olympic forests, and it is common from coastal areas well into the mountains. *(Keith Lazelle)*

Left top: The mossy bark of fallen trees provides excellent nurseries for seedlings. In the rain forest, where the dense growth of the forest floor is inhospitable to tree seedlings, nearly all trees get their start on nurse logs. *(Janis Burger)*

Left bottom: As nurse log seedlings grow into mature trees and the nurse log decomposes, the row of stilted trees that results is called a "colonnade." Colonnades are a common feature of Olympic's temperate rain forests. *(Keith Lazelle)*

Young hemlocks, old Douglas-fir. The thick, corky bark of mature Douglas-firs insulates them from most small wildfires, and seedling Doug-firs do well in bare ashen ground. Barring fires and other disturbances, the young hemlocks that grow up in the shade of old Douglas-firs will eventually dominate the stand. *(Janis Burger)*

Northern spotted owls have come to symbolize the old-growth forest ecosystem in the Northwest. Feeding on flying squirrels and nesting in cavities in snags, spotted owls need large tracts of old-growth forest to successfully reproduce. In the Olympics, spotted owls are near the northern extent of their range. *(Keith Lazelle)*

Pacific rhododendron flourishes in open Douglas-fir forests of the eastern Olympics, where it blooms from April in the lowlands through June in montane forests. An indicator of the relatively dry forests of the east Olympics, its range is opposite that of Oregon oxalis. *(Keith Lazelle)*

Right top: East Olympic rivers have a character and flavor all their own. Streams like the Duckabush are marked by narrow valleys and steep gradients. Down logs play an important role in these rivers, regulating high flows and creating pools and hiding cover for young salmon and trout. *(Keith Lazelle)*

Right bottom: The cool wet forests and shaded streams of the Olympics make ideal habitat for salamanders; eight species of salamanders and newts are found here. The western red-backed salamander is a small woodland species. Most woodland salamanders lay their eggs on dry land, and the young hatch as miniature adults. *(Janis Burger)*

Most elegant of the birds that feed on small fish and amphibians, great blue herons nest in colonies in dense forests and feed along the peninsula's streams, lakes, and coastal areas. *(Keith Lazelle)*

Spring brings brilliant flashes of color to swollen streams in the form of harlequin ducks. Harlequins leave coastal waters to mate and nest among snowmelt rapids of peninsula rivers. *(Keith Lazelle)*

One of the most spectacular parts of the outer coast is Point of the Arches. Sandstone strata at the south end of Shi Shi Beach fold into the complex assemblage of basalts, breccias, and conglomerates that form the Point. Rocks at Point of the Arches have been dated at more than twice the age of the oldest rocks in the interior. *(Keith Lazelle)*

Left: Cape Flattery at the northeast tip of the peninsula typifies the rugged, rocky shoreline of the northern coast. Short coastal streams bisect steep headlands, and intertidal organisms are subject to heavy pounding by winter seas. *(Keith Lazelle)*

One among several flowerlike sea anemones found in the intertidal area, aggregate sea anemones often form dense colonies. This species can reproduce sexually or asexually by cloning. Cloned colonies are known to maintain strict territorial boundaries. *(Keith Lazelle)*

Right top: Coralline algae and surfgrass. A common seaweed on the outer coast, surfgrass lines submerged tide pools and rocky shelves. Unlike almost every other seaweed (eelgrass is another exception) surfgrass is not an alga but a true grass with roots, flowerheads, and seeds. *(Keith Lazelle)*

Right bottom: Blood star on eelgrass. In more protected waters of the Strait of Juan de Fuca and Puget Sound, eelgrass provides important habitat for a variety of fish and invertebrates. The small blood star, like the brood star, retains its young, an unusual behavior for intertidal species. *(Janis Burger)*

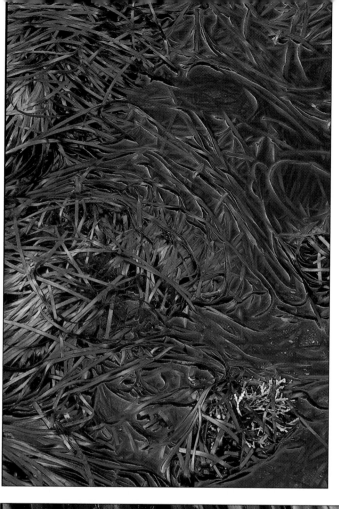

The elevational ranges of many intertidal species are determined by predation. Dense beds of California mussels and gooseneck barnacles come to an abrupt end where they meet the range of purple or ocher sea stars. Sea stars' aversion to drying sunlight limits their upward spread, allowing mussel beds above them to flourish. *(Olympic National Park)*

Gulls and seastacks near Cape Johnson. Olympic's offshore rocks and seastacks provide nesting habitat for more than 100,000 pairs of breeding seabirds. *(Keith Lazelle)*

Black oystercatcher nest. For many marine birds, a nest is a mere scrape on a ledge or amid the sod of offshore islands and seastacks. Oystercatchers' eggs blend well with the rock substrate. The eggs of common murres are pear-shaped to prevent their rolling off rocky ledges. *(Keith Lazelle)*

Right top: Though listed as threatened in Washington state, bald eagles are a frequent sight on the outer coast, where some 50 nesting pairs find habitat. In winter, eagles from northern regions migrate to the Olympic coast to feed in nearshore waters. *(Keith Lazelle)*

Right bottom: One of the most colorful of Olympic's nesting seabirds is the tufted puffin. About 23,000 puffins nest in burrows on islands off the outer coast; a few small colonies also nest on islands in Puget Sound. *(Olympic National Park)*

Although sea otters were extirpated from the Northwest coast by the early 1900s, their successful reintroduction in 1970 has resulted in a healthy breeding population in the park's coastal waters. Their luxuriant fur insulates them from cold North Pacific waters, but also makes them highly vulnerable to oil spills. *(Fred Felleman)*

Steller's sea lions visit Olympic Peninsula waters from northern breeding ranges. Larger and shaggier than California sea lions, the two mix amiably here, even sharing haulouts on offshore rocks. *(Fred Felleman)*

California gray whales often swim close to Olympic beaches straining bottom sediments through their fringed baleen as they migrate to their northern ranges. Once hunted nearly to extinction, gray whale populations now approach their original numbers. *(Fred Felleman)*

8
Seabirds and Mammals of the Outer Coast

South of Yellow Banks, on the northern Olympic coast, a small granite monument stands half hidden among surrounding trees. The Norwegian Memorial commemorates the three-masted schooner *Prince Arthur*, which ran aground in the winter of 1903, killing 18 of her 20 crew members. Farther south, a smaller stone marks the grave of 20 sailors who drowned when the Chilean schooner *W. J. Pirrie* struck an offshore island in the winter of 1920. This marker is not far from the site of another Chilean ship that wrecked here in 1883. The calls of gulls and slow lapping of waves are a quiet counterpoint to the image of ships cast up on the rocky shore.

The seacliffs, stacks, islands, and reefs of the Olympic coast have troubled the dreams of generations of mariners. Storm winds on the outer coast (as distinct from the more protected "inside waters" of Puget Sound and the Strait of Juan de Fuca) can exceed 150 miles per hour. Fog may blanket the coast for days, and safe anchorages are all but nonexistent. No one knows how many lives have been lost to this rugged coastline, but historical records list more than 150 ships that met their end here. Before 1875, some 60 Japanese junks are also known to have wrecked at various points along the Pacific coast. As recently as 1927, the Japanese fishing boat *Ryo Yei Maru* was found drifting toward shore south of Cape Flattery, still bearing the skeletal remains of its crew.

Hikers find pieces of shipwreck debris scattered among driftwood on coves and beaches. I have vivid memories of my first exploration of the coves north of Shi Shi Beach, where the wrecked World War II–era troopship *General M. C. Meigs* lies, its mast leaning sharply amid wave-pounded sea-stacks and billowing spray. Twenty-five years later the image lingers — its juxtaposition of human industry and the wildness of this untamed coast still poignant. But humans made their home here for thousands of years before the ships of exploration and commerce plied these coastal waters. And the jagged, rock-studded jaws of the coast so dreaded by mariners have provided a safe harbor for seabirds and sea mammals for many more millennia.

In the early years of this century, little was known to science about the seabird colonies nesting on the offshore rocks and islands of the Olympic coast. That changed in 1906 when an ornithologist named William Dawson first explored the area. Traveling by cedar canoe with his family and two Quileute guides, Dawson encountered one of the most productive seabird nesting areas on the Northwest coast. He documented rookeries of auklets, puffins, and storm-petrels and listed numerous other species using these offshore islets. His writings in scientific journals and the popular press helped build the support that led to President Theodore Roosevelt's executive order establishing migratory bird sanctuaries on the coast the following year.

Today, some 870 islands, sea-stacks, rocks, and reefs are protected as part of the Flattery Rocks, Quillayute Needles, and Copalis Rock national wildlife refuges, collectively known as the Washington Islands National Wildlife Refuges. The steep, rocky shores and inaccessible uplands of 36 major islands and hundreds of smaller rocks and reefs provide critical habitat for 80 percent of Washington state's nesting seabird population. The islands also provide habitat for threatened bird species that nest on the coast such as bald eagles and peregrine falcons. The islands' isolation offers protection from terrestrial predators, and their cliff faces, ledges, and glacier- and wind-deposited soils host a variety of nest sites. Some of the same factors that lead to the incredible productivity of Olympic's intertidal area — moderate climate, upwelling of deep-ocean nutrients, and undeveloped character — make this a prime area for seabirds.

Seabirds animate the rocky coast. Their bright spring plumage, their incessant calls, the vigorous trajectories of their flights charge the coastal atmosphere with life. They are the winged expression of the tremendous fecundity of the marine environment — feathered offspring of sunlight and sea. The sight of cormorants or oystercatchers skimming across the waves as I clamber over driftwood onto an open beach brings a special kind of gladness.

Bird Rocks of the Outer Coast

At first light in early summer, birds are already leaving the rocks and islands to fish. Auklets, guillemots, cormorants, and murres fan out over the waves and head for open water. Other birds more familiar to beach walkers peruse the tideline and seaweed-covered rocks: oystercatchers utter chattering calls as they careen low over the rocky shore, and ubiquitous glaucous-winged gulls patrol the tideline like cops on a beat. Overhead, where spruces trace the ragged edge of the sky, a pair of bald eagles circle on dark motionless wings while a gaggle of crows caw loudly and take turns diving at the eagles' broad white tails.

Along with bald eagles, other threatened and endangered species thrive along the coast. Marbled murrelets and peregrine falcons find refuge here; even brown pelicans have recently begun to visit the area in late summer and early fall. Sea mammals feed in productive coastal waters and haul out on rocky terraces and reefs. Sea otters continue to repopulate the coast, and Steller's sea lions, too, are increasing in number. Just offshore, the spring and fall migrations of California gray whales remind us of the long history humans share with sea mammals on the outer coast. With federal protection, California gray whales have made a remarkable recovery. Centuries after humans first arrived on these shores, the whales, seals, sea otters, and sea lions that return to the coast continue to stir the hearts of visitors.

Other cetaceans also ply the waters of the Olympic coast; harbor and Dall's porpoises, orcas or killer whales, and several species of dolphins, but few of these are easily seen from the shore.

The eagles and peregrines, shorebirds, seabirds, and marine mammals of the Olympic coast are part of the warp and woof, the tightly woven fabric of the Olympic ecosystem. By exploring the natural histories of some key breeding birds and resident and migratory sea mammals, we get a fuller sense of the coast's unique place within the ecosystem.

Recovery on the Coast: Bald Eagles and Peregrine Falcons

The old Sitka spruces that cling tenaciously to the tops of headlands seem to stand in defiance of the elements. Roots woven firmly into bedrock and stout limbs jutting into the wind, these aging sentries of the forest's edge are holdouts against the incessant advance of the sea. When at last they succumb, we find them weathered and broken, angled against a sea cliff or cast back up on the beach by waves, their tangled roots ground to nubbins. But while they stand — and some lean over the sea's edge for centuries — they serve as fitting perches for that largest bird of prey that hunts the coastline, the bald eagle.

Bald eagles *(Haliaeetus leucocephalus)* are strikingly noticeable along the coast all year. Their white head and tail feathers stand out in the misty light, and the dark silhouettes of their eight-foot wingspans are unmistakable against the sky. Eagles spend so much of their time perching conspicuously in headland spruces or gliding on thermals in search of food, they've come to share an irrevocable identity with this rocky coastline — so much so that for some, a visit to the outer coast seems incomplete without sighting one. Bald eagles lack the distinctive white head and tail feathers for their first four years, but their shape is hard to miss. Golden eagles, with which they are sometimes confused, inhabit the interior Olympics and are much fewer in number.

Bald eagles begin their courtship behavior as early as December in some years, and their displays can be quite spectacular. Pairs will sometimes lock talons and plummet from heights in a series of somersaults, or spiral upward close together on soaring thermals. They usually nest in March or April, constructing or refurbishing bulky platforms of large sticks and lining them with finer material. Not all these nesting materials are gathered from the ground. One biologist observed a bald eagle descend as if to land on a stout limb that jutted from the dead top of an old Douglas-fir. But rather than slowing as it approached the tree, the bird maintained speed, stretched out its talons and hit the limb hard, snapping it off with a riflelike crack. The eagle then flew off with the large limb grasped firmly in its talons.

Eagles seem to prefer rocky headlands, islands, even some sea-stacks over beaches for nesting sites. Eagles nesting on the open coast seldom use the tops of trees, as they do in inland areas. Instead the nests are usually found farther down in the canopy of tall Sitka spruces — a concession no doubt to Pacific storms. Nests are maintained and repaired each year, but only one nest in a given breeding territory is occupied. Eaglets are born in late April or May and usually fledge in mid-July. Their early weeks are spent close to the nest site, learning to hunt and scavenge. Dead and dying fish are generally the staples of the bald eagle's diet, but some biologists suspect that eagles on the Olympic coast may rely more heavily on birds. Having watched eagles casually pluck gulls from the waves, I believe this.

A curious migration occurs throughout Washington's coastal waters in late summer. Newly fledged eagles and their parents disperse for a few months in the fall, returning again in December. Telemetry work has shown that some Washington birds fly north to British Columbia to feed on early-returning salmon runs. It may be that this strategy allows home territories to recover from the intense feeding associated with nesting, or the birds may need the added nutrients of a salmon diet to store up energy for winter. In early winter, a much larger migration occurs. Bald eagles from as far away as Alaska, interior British Columbia, and the northern Rocky Mountains flock to Washington's coastal areas as inland waters freeze up. Large numbers of these overwintering eagles from far-flung territories mingle with Washington birds along coastal and inland waters, putting aside the territorial behavior they exhibit so robustly during the nesting season.

For the past several decades the remote forested coastlines of Washington, British Columbia, and southeast Alaska have served as a stronghold of North America's bald eagles. Here, feeding on plentiful fish from the Pacific and shorebirds and small mammals on the uplands, they were able to persist while bald eagle numbers elsewhere declined sharply. Pollutants in lakes, rivers, and coastal waters, widespread declines in fish runs, and destruction of nesting and roosting habitat all took a toll.

Eagles are still listed as threatened throughout most of their former range in the United States, but with protection under the Endangered Species Act and the banning of DDT and other harmful pesticides, they have made a strong comeback in the past two decades. Although listed as threatened in seven Pacific states, bald eagle numbers in Washington are approaching (and in some cases exceeding) target figures established in the U.S. Fish and Wildlife Service's Pacific Bald Eagle Recovery Plan. Nesting success throughout Washington has shown a steady upturn during the 1980s, going from 86 young produced in 1975 to more than 400 in the early 1990s. In 1994, there were 450 occupied nests in Washington state, more than 200 of them on the Olympic Peninsula.

Although some of this increase may be due to better observation, there is no doubt that bald eagles are doing well here. Twice-yearly surveys of bald eagles on the Olympic coast have identified some 53 nesting territories north of the Queets River. In 1994, 47 of these were occupied by nesting pairs, with 49 young counted in June. This success undoubtedly reflects the quality of the coastal habitat, but surveys of more developed areas in the Strait of Juan de Fuca and Puget Sound show that the number of nesting territories has increased there as well. Unfortunately, the small number of bald eagles that nest along Olympic Peninsula rivers and those nesting along Hood Canal on the eastern shore of the peninsula are often unsuccessful in raising young. Some biologists suspect this may be due to declining salmon runs and pollutants along the canal. On the Columbia River, where salmon numbers have been decimated by dams and pollution, very few chicks hatched. Although the overall picture is improving, other states in the Pacific recovery area still have a long way to go before their eagle populations are considered healthy.

Another bird of prey that nests along the Olympic coast — though one seen far less frequently than bald eagles — is that living incarnation of flight called the peregrine falcon *(Falco peregrinus)*. Renowned for their speed and mastery of the air as hunters, peregrines were revered in ancient Egypt and coveted by feudal lords. Many ornithologists today consider peregrines to be among the most highly evolved birds of prey. Inhabiting the wildest, least-disturbed places — desert ranges, canyons, undeveloped coasts — and nesting on steep, inaccessible cliffs, peregrines at one time enjoyed the widest distribution of any bird in the world. They feed exclusively on live birds and take their prey on the wing.

Dropping into a stoop from a mile above the earth, they attain speeds of more than 200 miles per hour, yet control their flight with pinpoint accuracy. Hunting at sea, peregrines are known to cruise low in the troughs of rolling waves to surprise unsuspecting seabirds.

Historically, thousands of peregrines followed the migrations of other birds along the great flyways of North America. Yet by the mid-1960s, fewer than 20 known pairs of peregrine falcons remained in all of North America. The presence of DDT and other pesticides in the environment resulted in thin, brittle egg shells and pushed these magnificent birds to the point of extinction. DDT was finally banned in the United States in 1972. Since then, an intensive captive breeding effort has allowed peregrines a remarkable recovery. Captive-bred birds and their offspring now populate former ranges in the eastern United States, the Rocky Mountains, and much of the West. Although no captive-bred falcons have been introduced on the Olympic coast, some 14 peregrine nesting territories have been identified here. In 1994, 12 of these were occupied, producing 17 young. With this rate of nesting success, the year-round presence of falcons on the Olympic coast seems secure.

Olympic's peregrines nest in scrapes on steep sea cliffs as well as off-shore sea-stacks. They feed primarily on smaller seabirds: auklets, storm-petrels, guillemots, and puffins, but like peregrines elsewhere, most any bird will do, including an occasional crow or robin. They also hunt Bonaparte's gulls, and bufflehead ducks, taking on the average one bird a day. Falcon pairs roost together and hunt cooperatively. Since peregrines usually hunt offshore, and much of their time is spent roosting (where they blend in well with the dark canopies of trees), sightings are uncommon. Only a few other nests have been reported in Washington, in the San Juan Islands of Puget Sound and the Columbia River gorge.

Nonresident peregrines can sometimes be seen migrating along the coast, particularly in spring. Migratory birds winter along the Gulf Coast, through Mexico and Central America and as far south as Chile and Uruguay. Since DDT is still used in some of these countries, the fate of North American peregrines is still far from secure. As recently as the late 1980s falcons nesting along the Big Sur coast of California were still suffering nesting failures due to thin eggshells. Such instances should make us all the more protective of those remaining wild enclaves like the Olympic coast, which have allowed these resident masters of the wind to thrive.

Bird Life on the Rocky Shore

If there is a single bird that embodies the wave-swept intertidal area of the Olympic coast, it is the black oystercatcher *(Haematopus bachmani)*. Its sharp, whistled, staccato notes are synonymous with the sound of waves rolling over rocky reefs. Its dark plumage, set off by a bright red bill and ringed yellow-gold eyes, suggests the hidden colors of a coastal tide pool. Oystercatchers seem as adept and well suited to their tumultuous intertidal environment as dippers are to mountain streams. I've watched them feed on rocky reefs all but oblivious to the waves breaking around them. A last-minute hop or skip sets them effortlessly clear of harm's way. When seriously involved with a mussel or chiton, they often just let small waves wash over them. As backwash streams off the rocks, they reappear, still occupied with their meal, as if the surf were no more cause for alarm than a passing shower.

Oystercatchers are superbly adapted to their intertidal world. Their legs and feet are tawny and strong enough to grip the substrate, and their long beaks are stout and effective. In spite of their name, black oystercatchers eat very few oysters, but they have little trouble popping limpets from the rocks and working them free of their shells. Chitons, small crabs, even mussels offer little challenge. Working the surf line where mussels often remain at least partly open, oystercatchers cut their quarry's abductor muscles with quick stabs from their beaks. A few shakes then frees the mussel from its shell. Small crabs or tightly closed mussels require a bit more effort, but these are pummeled open with the chisellike tip of the oystercatcher's beak and consumed without further fuss. The effectiveness of these birds as feeders is apparent in the low numbers of limpets in feeding territories occupied by the large black birds. Few other birds besides gulls and occasional crows compete with oystercatchers for large intertidal mollusks. Even so, oystercatchers fiercely defend their territories well beyond the nesting season.

Black oystercatchers locate their nests in depressions on rocky ledges, sometimes behind clumps of sod, on sea cliffs or islands (Destruction Island hosts large numbers of them). Some are loosely lined with pebbles or shells. Both males and females incubate eggs and feed their young, often far beyond fledging. Young oystercatchers, which are a lighter brown than their parents, must learn the tricks of foraging before striking out on their own. Ravens and gulls are the primary predators of eggs and chicks, but the parent birds are remarkably effective in fending them off. The old broken-wing ploy is often trotted out, and more inventive games resembling hide-and-seek have been reported.

Oystercatchers seem unafraid to nest in glaucous-winged gull colonies, and there may be some advantages in this. Gull nests may signal that a site is safe from predatory mammals; gulls may serve as an early-warning system when danger is about; and intertidal invertebrates may be more plentiful as a result of nutrients from gull wastes.

Oystercatchers are a common sight on rocky coasts from California to Alaska. They inhabit the Olympic coast year-round and are probably joined in winter by migrants from the north. Some 90 nests were identified along this coast in the 1980s, producing more than 300 birds. Far fewer inhabit the shores of Puget Sound, and oystercatchers have disappeared entirely from its southern waters. But their rhythmic calls and bright good looks continue to animate the open rocky coast.

The black oystercatcher is easily recognized by its stout red beak. These shorebirds feed on shellfish and crustaceans on exposed rocks of the intertidal zone and utter sharp chattering calls as they fly above the surf.

Wherever oystercatchers are foraging, off-season visitors are likely to see two smaller shorebirds feeding among

Janis Burger

them. Black turnstones and surfbirds are robin-sized shorebirds that migrate along the Olympic coast. Common in fall and winter, flocks of black turnstones frequently join oystercatchers on reefs at low tide, searching among rockweeds and kelp for small limpets, barnacles, and other crustaceans (their cousins, the ruddy turnstones, are abundant in spring). A stately combination of black and white, their upper parts becoming striped in spring, turnstones can be seen probing cracks and fissures in the rocks and poking about beneath them for food. True to their names, they do turn rocks over — some of them quite sizable — with a quick flip of their bill. When the tide is in, turnstones also rummage through seaweed litter on the beach (sanderlings sometimes follow behind them, feeding in the freshly exposed sand and gravel). Gregariously social and keeping up a constant chatter as they feed, black turnstones are occasionally joined by smaller numbers of surfbirds.

These intertidal feeders closely resemble turnstones, but they are slightly larger and a lighter gray. In spring, surfbirds turn a beautiful pattern of brown, black, and white. Both are inland breeders (surfbirds migrate north in summer to nest in Arctic tundra). But in fall, winter, and early spring, they feed in flocks along the coast. A flock of 100 to 200 is known to winter on Destruction Island. I sometimes see them hopping from rock to rock in continual search for tiny shellfish, which they swallow in the shell. Like turnstones and most other shorebirds, surfbirds consume about a third of their body weight in food each day. Interestingly, though black oystercatchers rigorously protect their territories from others of their kind, including dispersing young, they show a marked tolerance for these migratory neighbors. One of the most appropriately named of shorebirds, the surfbird's genus name, *Aphriza*, means "lives in the sea's foam."

Cormorants, Puffins, and Auklets

Much like the intertidal habitats of the Olympic coast, the offshore rocks and islands are located at an interface between northern and southern seabird nesting ranges. As a result, the outer coast of Washington supports a greater diversity of species than coastal areas to the south, but the total number of birds nesting here is lower than either Oregon or British

Columbia. Though the Washington Islands refuges harbor some 110,000 pairs of nesting seabirds, the rocks and islands lie at the edge of both northern and southern ranges for many species, and a number of species do not breed here. The coastal islands do serve as important nesting habitat for some 14 species of seabirds and are critical for a few species, such as the rhinoceros auklet. Seabirds spend much of their time feeding out on the open water. But among the species most distinctive and easily recognized by visitors on the shore are the cormorants.

Three species of cormorants nest on the outer coast. All are large, dark, long-necked seabirds that perch upright with graceful S-shaped curves of their necks and lifted beaks. In their trademark posture, they perch with wings held loosely out to their sides: cormorants lack ducks' water-repellent plumage, and must frequently spread their wings to dry. Cormorants have serrated, hooked bills like those of mergansers (excellent for catching and holding fish), and they store food in pouches below their lower mandibles. Their wide webbed feet, like those of diving ducks, are designed to maximize underwater propulsion.

Pelagic cormorants *(Phalacrocorax pelagicus)* are the smallest of the three and by far the most common. They differ from the others by their dull red faces and throat patches and by white patches on their flanks during the spring breeding season. Pelagics dive quickly, without a splash, and swim swiftly underwater in pursuit of herring, sand lance, sculpin, and other fish. Pelagic cormorants breed throughout Puget Sound as well as on the outer coast. The number of breeding pairs on the coast increased from around 700 in the mid-1980s to more than 2,000 by 1992. In 1994, the number of breeding pairs dropped back to 1,500, probably because of warm surface water and changing fish populations associated with El Niño. El Niño or southern oscillation events occur regularly off the coast of South America, triggered by disturbances in the pattern of trade winds. As warm water flows up the coast, it changes the pattern of currents and hampers the upwelling of deepwater nutrients onto the continental shelf. This, in turn, limits the productivity of phytoplankton and other marine organisms (that form the basis of the coastal food chain) and sends a ripple effect through the entire marine ecosystem, affecting top-feeders like salmon, seabirds, and marine mammals. Some species along the coast have reacted dramatically to El Niño events.

Double-crested cormorants *(P. auritus)* are also a frequent sight on the coast. They can be distinguished from the others by their orange throat pouches (the double head crest is usually not apparent). In spring their black plumage takes on an iridescent green sheen. Double-crested cormorants are a northern species, finding their southernmost breeding habitat on Washington's coastal islands. Some reside on the coast year-round, while others migrate to the more protected waters of the Strait of Juan de Fuca and Puget Sound in winter. In good breeding years the coastal islands host 400 to 800 breeding pairs; in El Niño years, numbers may drop as low as 200.

Brandt's cormorants *(P. penicillatus)* are least numerous of the three. Brandt's is a southern species; the Olympic Peninsula and Vancouver Island are the northern extent of its range. During breeding season, Brandt's is distinguished by a blue throat pouch with a yellow patch behind. Only one breeding colony is known along the coast; between 100 and 200 pairs of Brandt's cormorants nest on Destruction Island, often in association with glaucous-winged gulls. Many also winter in Puget Sound. When disturbed, all three cormorants will fly away from their nests, leaving their young exposed to predators.

Three similar species nesting together and sharing the same food resources would be expected to have evolved some noncompetitive breeding strategies — and they have. The three nest at different times. Double-crested cormorants nest earliest, usually by early May. Pelagic cormorants come next, then Brandt's cormorants begin nesting in late May. Nesting locations are also somewhat stratified. Double-cresteds prefer to build their messy, guano-covered nests on the upper slopes of offshore islands; pelagic cormorants prefer steeper, rocky or sand cliff faces, usually below them. Brandt's cormorants prefer the tops of islands as well as south-facing slopes. The staggered nesting times and locations may minimize competition when feeding nestlings. It's possible that cormorants divide up some some food resources as well.

Tufted puffins *(Lunda cirrhata)* are among the most charismatic of Olympic's offshore breeding birds. They are seldom seen close to shore but are well known to boaters and fishermen — from whom they rob herring off baited lines quite effectively. In summer plumage their clownish faces sport oversized multicolored bills and long straw-colored tufts that curl shaggily behind their eyes. The tufted puffin's face is white, the rest of its body jet black. Like other alcids of the coast — guillemots, murrelets, auklets, and murres — the puffin uses its short powerful wings to "fly" underwater in pursuit of fish. Its large beak proves handy in carrying numbers of fish back to its young, and it can hold fish securely in its beak with its tongue while continuing to catch more. Herring, smelt, sardines, and perch are among the puffin's preferred prey.

Tufted puffins are common from north of the Aleutians to British Columbia, and they are localized residents from Washington to Baja. They nest on islands of the outer coast; a few small colonies also nest on islands in Puget Sound. Puffins nest in burrows in steep turf-covered slopes or cliff tops; they seem to prefer high locations for easy takeoffs. Their burrows are commonly two to six feet long and five to six inches in diameter. A single egg is laid on bare ground in a cavity at the end.

About 23,000 tufted puffins are known to nest in 25 colonies along the Olympic coast. The largest colonies are on Alexander and Carroll islands. As recently as the 1950s and 1960s, more than 1,000 tufted puffins also bred on islands in Puget Sound; Protection and Smith islands each harbored more than 300 birds. Today, only 13 pairs remain on Protection Island. Breeding pairs elsewhere on inside waters have been wiped out by agriculture and residential development, predation by pets, and indiscriminate shooting. Puffins may be among the most disturbance-sensitive of all nesting seabirds. They seem exceptionally high-strung, leaving their nests frequently when trespassers are near, keeping an eye on them, flying over, and harrying them.

Fortunately, puffins' breeding habitat on the outer coast is safe from most such disturbances, and large breeding colonies in British Columbia and Southeast Alaska number in the hundreds of thousands. In an attempt to restore puffins to Protection Island — now a U.S. Fish and Wildlife Service refuge — biologists plan to introduce puffin chicks to the island and feed them artificially until they fledge out. Similar efforts have proven successful with puffins on the Atlantic seaboard, and scientists hope that a 10- to 12-year restoration effort will rebuild populations on this recently protected island refuge. In the meantime, tufted puffins do make appearances on the Strait of Juan de Fuca and Puget Sound when they leave the outer coast in winter. Others head out to open sea or to slightly warmer waters to the south.

An alcid more commonly seen in the waters of the Olympic coast is

Janis Burger

The rhinoceros auklet, a common seabird along the coast, nests in burrows on offshore islands. Protection Island in Puget Sound and Destruction Island on the Pacific coast are among its major breeding grounds.

the rhinoceros auklet *(Cerorhinca monocerata)*. Described by some as a "misnamed puffin," they are dark like puffins but with a lighter breast. Rhinos are named for the small stout horns that sprout from their upper mandibles during breeding season; they also wear white "eyebrows" and understripes then. Like puffins they nest in burrows on offshore islands but leave to find food and return to their nests in the predawn and after dark. The peninsula's offshore islands form a major part of the rhino's breeding territory. Colonies of about 13,000 birds nest on Destruction Island, their largest southernmost breeding colony; another 17,000 breed on Protection Island in Puget Sound. Rhinoceros auklets are excellent flyers; birds fledging from Protection Island fly to the outer coast to feed. Like puffins, most rhinos also spend their winters at sea.

Storm-petrels, Murres, and Gulls

Among the more far-ranging birds that nest on the outer rocks are the storm-petrels. These small, songbird-sized flyers range from the Arctic to the Galapagos Islands. Storm-petrels do not dive, but skim the ocean surface for fish, small squid, and crustaceans. Fork-tailed storm-petrels *(Oceanodroma furcata)*, which are mostly gray, breed on offshore islets from Oregon to Alaska. Closely related Leach's storm-petrels *(O. leucorhoa)*, which are black with white rump patches, find most of their home breeding range on rocks and islands off the Olympic coast. Both birds spend most of their time on the water, coming ashore only to breed. When William Dawson explored the Olympic coast in the early years of this century, he found an island off La Push "honeycombed" with the shallow burrows of Leach's storm-petrels and estimated a Washington population of 100,000 birds. More recent estimates set the population at 36,000 for Leach's, around 4,000 for fork-tailed storm-petrels.

During breeding season, storm-petrels are entirely nocturnal, leaving their nests and returning under cover of darkness. Ill-suited for getting around on land and small enough to be ideal prey for peregrine falcons as well as gulls, river otters, and other predators, they are even reluctant to venture out on moonlit nights. Ornithologist E. A. Kitchen, who studied birds on the peninsula in the 1930s and 1940s, wondered how the fork-tailed storm-petrel, which he called "this mite of a bird," is able to survive the storms of the open ocean where it spends the nonbreeding months. Fork-tails are more pelagic than Leach's storm-petrels; during nesting they feed much farther out at sea. Consequently, their eggs and chicks are well adapted to being left for days at a time.

Leach's storm-petrels are known for their dense colonies; burrows on one islet in the Quillayute Needles average 12 holes per square yard. Ecologists have long proposed that herding behavior is in the best interests of individuals of many species because it lessens their chances of predation. The same may be true for nesting seabirds. It may be that the nesting behavior of the Leach's storm-petrels of the Olympic coast developed in the presence of a predator, though there is no major predatory threat there now.

Closer to shore, one of the most frequently seen of Olympic's nesting shorebirds is the common murre *(Uria aalge)*. Murres are easily identified by their compact bodies, dark backs, white sides, and long slender bills. They nest in crowded colonies on offshore rocks and islets, their "nests" mere scrapes on ledges and clefts of rock. During the spring breeding season, I've seen them from shore, sitting upright almost shoulder to shoulder, penguinlike, along a high rocky ledge. Their single, pear-shaped eggs are well suited for staying put on narrow ledges. Murres are the largest and probably most commonly seen seabirds along the coast year-round. They nest strictly on the outer coast. In winter, they congregate in large flocks and are frequently seen in the waters of the Strait and Puget Sound. At rest on water, they tilt their beaks up, a good identification point.

Like other alcids, murres are excellent divers and powerful underwater swimmers. They can typically dive 100 feet deep in pursuit of fish, and are capable of reaching depths of 500 feet. Alcids are as well adapted to underwater swimming as to flight, and murres take large numbers of sand lances, smelt, herring, and some bottom fish. But unlike some other alcids, they can carry fish to their young only one at a time. Murres frequently locate their colonies adjacent to double-crested, pelagic, and Brandt's cormorant rookeries, an arrangement which, like cormorants and gulls, may provide some advantages for murres.

Historically, common murres were quite numerous off the Olympic coast. Before 1982, the coastal population numbered from 20,000 to 30,000 nesting birds. In 1983 the population crashed to one tenth of that, and it has remained depressed ever since. Research done at the University of Washington suggests that El Niño conditions triggered the population drop in 1983. Although this has no doubt occurred several times in the past, murre numbers have traditionally recovered following a return to normal conditions. This decline is particularly disturbing for its persistence. Studies conducted on Tatoosh Island suggest that the recovery of bald eagles on the coast may be a contributing factor. Bald eagles flying over murre and cormorant colonies cause nesting birds to flee, thereby allowing gulls to prey on murre eggs and chicks. Ulrich Wilson, a biologist with the U.S. Fish and Wildlife Service, points out that once a seabird population has been drastically reduced, disturbances that could normally be sustained by a larger population can be seriously detrimental to the remnant populations.

A study conducted by Wilson on the effects of El Niño events on common murres as well as double-crested and Brandt's cormorants on the Washington coast found that their numbers decreased markedly during El Niño occurrences in 1982–83 and 1987. But Niño's effects on murres were complicated by several types of human-caused disturbance.

Ruth Kirk

Tens of thousands of murres and other seabirds were killed in each of several oil spills, and thousands more murres became entangled in commercial gill nets (about 2,500 were killed this way during the 1989 nesting season alone), events that may have conspired to keep murre populations depressed. Other human-caused activities — including the U.S. Navy's practice bombing of the wrong island — certainly took a toll as well. In general, encroachment of humans and their animals on breeding islands and overfishing of seabird prey species have had severe impacts on seabird populations along the entire west coast of North America. These are further compounded in areas affected by El Niño events. Along with understanding the environmental factors at work on seabird populations, the future conservation of seabirds requires that we fully comprehend and manage our own impacts as well.

Common murres, the most abundant of Olympic's breeding seabirds, nest in dense colonies on steep offshore rocks. In winter, they flock together on open water.

One group of seabirds that doesn't seem to be suffering at all on our account is the gulls. Both western and glaucous-winged gulls nest on the coastal islands, while a number of other species migrate through or winter here. Glaucous-winged are the most commonly seen gulls along Olympic Peninsula shorelines at any time of year. They tend to favor the inside waters of the Strait and Puget Sound; Protection Island alone harbors a breeding colony of over 3,000 nesting pairs. Western gulls are fewer in number and prefer the outer coast from Washington south to California; Destruction Island is one of their northernmost nesting colonies. But their natural history becomes somewhat complicated on the outer coast because the two species interbreed. In general, glaucous-winged gulls are distinguished by their white heads and pale gray plumage; western gulls have dark gray mantles. Both species carry yellow bills with distinctive red spots.

Maybe it's their strident voices, or their propensity for feeding at garbage dumps and sewage treatment plants, but gulls rank low on most bird-lovers' appreciation lists, and that is unfortunate. As Kitchen has pointed out in his adulatory style, "Few birds are their equal in gliding through the air, either with or against the wind. They are wonderful swimmers and many varieties are expert divers. . . Gulls are at home on land, walking or running on the sand or a hard surface. Few birds can do all these things well." Gulls do seem to enjoy loafing about the beach or holding court along rooflines in Neah Bay and La Push — maybe that's why I'm drawn to them.

Both gull species are serious predators on other seabird eggs and chicks, going after auklets, puffins, murres, cormorants, guillemots, and others in season. They've learned to drop shellfish on rocks to crack them open (a trick they might have stolen from crows), pirate the catches of a number of diving birds, forage over all marine environments, and scavenge virtually everywhere. One does not encounter a high level of concern for gulls among most biologists. As one researcher put it, "Gulls are very smart and opportunistic; they'll take care of themselves. They may well be the last seabirds left." So call me sentimental, but the sight of a flock of western gulls rising and circling on the oncoming edge of a Pacific storm is a piece of the coastal world that still brings a smile.

One bright spot on the coastal horizon has been the recovery of the brown pelican. Almost pterodactyl-like in appearance, this large, shovel-beaked, and somewhat ungainly bird flies with heavy wingbeats and dives with wings extended, almost crashing into the water. A special tissue beneath the skin cushions the impact and pops the bird almost instantly back to the surface — usually with a fish in its beak. Like the peregrine falcon, brown pelican numbers were drastically reduced by pesticides and habitat loss. The only viable populations in North America remained in Florida. The ban on DDT has allowed them to recover in California and Mexico as well as along the southeast U.S. and Gulf coasts, and populations there seem to have stabilized. As El Niño events shifted the pelicans' prey populations northward, they have been showing up along the Washington coast in late summer and staying through fall. The first sightings were associated with the 1982–83 El Niño, but annual visits persist. A recent fall survey off the Olympic coast identified between 1,100 and 1,200 birds. Grays Harbor and Willapa Bay in southwest Washington reported thousands.

Pelicans are not often seen from shore this far north; they are among the many species that dive and fish in offshore waters. But another family of migratory animals feeds close in to the Olympic shore; some even haul out on nearshore rocks and islands. Visible from beaches, headlands, and bluffs, sea mammals tie this rocky coast to offshore waters for thousands of miles north and south.

Mammals of the Coastal Waters

The warm days of early spring bring the first blooms to coastal forests. Crimson salmonberry blossoms brighten trailsides, and pioneer violets wink among mosses and newly greening leaves. The large yellow bracts of skunk cabbage fairly shout from marshy spots along the streams and seeps, and swelling knots of coltsfoot erupt through roadside gravel. Spring brings an equally dramatic bloom of marine plants and animals to nearshore waters, and a succession of migratory sea mammals is drawn to the Olympic coast in response. By late spring, foraging sea lions can be seen (or heard) hauled out on offshore rocks and islands. Northern fur seals and elephant seals fish the offshore waters, along with the large resident population of harbor seals. Sea otters surface and bob lazily among kelp beds, meticulously grooming their handsome fur. And in March and April, one of the greatest migrations on the planet skirts the

Olympic coast as California gray whales swim close to shore on their 6,000-mile annual migration between their summer and winter feeding grounds.

Migrations of the Grays

Gray whales are true harbingers of the turn of seasons on the coast, and each spring I make a pilgrimage with family and friends out to greet them. From an overlook on a favorite headland, we watch for the spouting plumes that lift briefly into the wind and vanish as these gentle, 30- to 40-ton mammals sound and dive in the offshore swells. From their calving grounds in Baja's coastal lagoons to summer feeding grounds far north in the Chukchi and Bering seas, migratory grays form a season-long procession on the outer coast. Pregnant females and adult males are the first to arrive here, their knobby, finless backs cresting above the waves. They are followed by juveniles of both sexes. Toward late spring and summer, mothers and their calves can be seen close to shore, feeding among the kelp beds. Gray whales *(Eschrichtius robustus)* follow the contours of the continental shelf over the course of the spring, feeding on plankton, small schooling fish, and shrimplike organisms called krill.

Grays, like other baleen whales (humpback, minke, sei, fin, and right whales), strain their food through the fringed baleen that hangs from their upper jaws. Grays are primarily bottom feeders, straining tons of sediments as they feed. Some grays enter the Strait of Juan de Fuca, staying and feeding for a few days or weeks; some reside here year-round. Many more stop to feed in shallow bottom sediments off the coastal mouths of rivers like the Quillayute and Hoh. This is the best chance for coastal visitors to catch something of the whales' immense grace as they dive and surface, their great fluted tails flashing balletlike in the churning water.

Decimated by commercial hunting in the 19th century, gray whales were eliminated from the Atlantic; Pacific populations were reduced to a few thousand. But California gray whales have made an astounding recovery since they were listed under the Endangered Species Act and protected by the International Whaling Commission in the 1970s. Today the Pacific coast population exceeds 20,000, and most biologists consider gray whale numbers to have returned nearly to their original level. The grays were removed from the endangered list in 1994. Unfortunately, this isn't true for other small-toothed cetaceans that ply Washington's coastal waters. Minke, humpback, fin, and other whale populations remain small by comparison. Perhaps most threatened are right whales; their population may be down to no more than 50 to 200 animals.

Gray whales and other sea mammals were important sources of food and materials for Native American hunters who lived along the Olympic coast. The archaeological record, along with a large body of myths and stories, reflects the importance of these creatures to the material and spiritual fabric of early coastal life. The Makah and Quileute were whaling people, and much of these tribes' social structure centered around this demanding and highly skilled way of life. This type of subsistence hunting, even as practiced by the Makahs, who excelled at taking sea mammals, probably had little effect on the abundance of sea mammals at the time. But human impacts on the marine mammals of the North Pacific accelerated rapidly with the arrival of Europeans.

Following the visits of Vitus Bering in 1741 and James Cook in 1778, reports of the wealth of fur-bearing sea mammals spawned an aggressive trade in furs. Over the next century, the killing of countless numbers of seals, sea lions, sea otters, and whales for fur, baleen, oil, and meat drove many species to the brink of extinction. By the late 19th century, most of the species commercially hunted had reached "economic extinction," a point at which their numbers had plummeted so low as to make hunting them unprofitable. International treaties in the first years of the 20th century, as well as federal statutes and the removal of bounties in more recent decades, allowed most species to recover somewhat. A few, like California gray whales and sea otters, now approach pre-exploitation numbers. Others, like the blue and right whales, continue to hang on by a thread.

From the distant plumes of gray whales sounding past the breakers to the simple grace of a harbor seal surfacing in a nearshore swell, the Olympic coast offers a unique opportunity to encounter marine mammals in their natural environment. The story of how these animals interact with this coastal landscape helps make the offshore rocks and tangled kelp beds come to life for us.

Harbor Seals and Elephant Seals

Few marine mammal species breed and reside year-round off the Olympic coast; winter seas are prohibitive. But located at the interface of northern and southern breeding ranges for a number of species, these productive coastal waters support a wealth of seasonally foraging sea mammals. Among the few year-round residents, harbor seals are by far the most plentiful, as well as the easiest to observe. I often notice a dark, round, earless shape bobbing low among the kelp beds, large eyes turned to the beach. When I look again, it has dropped out of sight, only to reappear — and look back — a little farther out.

Harbor seals *(Phoca vitulina)* will often forage quite close to shore at high tide. At low tide, groups will haul out on low rocks or reefs, never far from the edge of water. Viewed from a nearby headland, their sleek, silvery or dark gray pelage is usually covered with whitish spots or rings. Like other pinnipeds ("fin-footed" mammals), seals evolved paddlelike hind flippers which make them swift and graceful swimmers under-

Among the few sea mammals that breed in Olympic waters, harbor seals can be seen year-round along most Olympic coasts. Although they were formerly hunted for bounty in Washington, they are now protected, and their numbers have been steadily increasing in recent years.

Ruth Kirk

water. They can dive to depths of 600 feet and remain submerged for 20 minutes — though three to five minutes is more common. On dry land, however, seals are awkward and clumsy. Pinnipeds took to the sea about 20 million years ago, much later than the cetaceans — the whales, dolphins, and porpoises — and must come ashore to rest, breed, and give birth to their young.

Harbor seals pup in May and June on the outer coast, a good bit earlier than the seals of Puget Sound and Hood Canal. Females nurse and tend to their pups continuously for a month to six weeks, then teach them to fish. Females occasionally leave their pups on the beach during this time; most of these pups are not abandoned, as visitors sometimes think, and they should always be left undisturbed.

Harbor seals are fairly opportunistic feeders, taking whatever is available, usually bottom and middle fish — herring, pollock, hake, and such. For years it was mistakenly believed that harbor seals fed primarily on salmon. From the 1920s through the early 1960s, the Washington Department of Fisheries offered a bounty on seals. In the years between 1947 to 1960, the state paid bounties on more than 17,000 animals without apparent benefit to commercial fisheries. In fact, harbor seals take relatively few salmon; salmonids are generally faster swimmers than seals, and most mature salmon pass through seal habitat only on their return to their rivers to spawn.

The Marine Mammals Protection Act in 1973 has allowed the number of harbor seals to increase considerably in Washington; seals on the outer coast are increasing at about 8 to 10 percent each year. Of a statewide population of between 30,000 and 40,000 in 1992, about 5,000 were found on the Olympic coast, distributed among some 50 haul-out sites. The more protected estuaries of Grays Harbor and Willapa Bay to the south support considerably larger numbers. Harbor seals have few predators on the coast, though orcas and sharks take a few pups. And since they feed on whatever is abundant, it is expected that their numbers will continue to increase.

Warm weather also brings much larger elephant seals up from their California and Mexico breeding ranges to forage in Olympic's coastal waters. Northern elephant seals *(Mirounga angustirostris)* are large; males weigh up to 8,000 pounds. Females are much smaller but can still reach 1,000 pounds. The males, which are our usual migrants, have bulbous protruding snouts (thus their name), and are known to float lazily on the surface, resembling nothing so much as large sunken logs. They come ashore rarely, to breed in their southern ranges and occasionally on the Olympic coast in summer to molt, a unique trait among seals.

Elephant seals are phenomenal divers and can pursue their prey of rockfish, squid, rays, and other deep-water fish to depths of more than 2,500 feet. They actually spend more time underwater than on the surface — behavior more often associated with cetaceans like sperm whales than pinnipeds. Even so, elephant seals must have proven easy targets for commercial sealers when they came to the surface to nap, for by the mid-1880s they had nearly become extinct. In less than a century, a breeding population numbering in the hundreds of thousands was reduced to fewer than 100 animals isolated on a group of islands off Baja. Since then their numbers have recovered somewhat, and for the past 20 years elephant seals have been expanding their range to the north, reoccupying former breeding sites. They recently recolonized a breeding site off the central Oregon coast, and frequent sightings by fishermen off Washington's outer coast and in the Strait of Juan de Fuca are promising. Unfortunately, the recovery of northern fur seals has been less dramatic.

Fur Seals and Sea Lions

Most pelagic of the seals, northern fur seals *(Callorhinus ursinus)* spend the greatest part of their lives in open water, where they sleep curled on the surface. They tend to favor the outer edge of the continental shelf to feed. At the peninsula's northwest tip, the Juan de Fuca canyon brings the shelf edge close to Cape Flattery, and fur seals figured prominently in the lives of the Makah people who still inhabit the cape. Not as large as elephant seals, male fur seals can reach 600 pounds. Females, which more often show up off the Olympic coast, are considerably smaller. Champion swimmers, they migrate from their breeding grounds in Alaska's Bering Sea as far south as California. But their numbers are not what they were.

Fur seals were decimated by a century and a half of unrestricted commercial hunting, during which sealers took breeding females as well as males. A 1911 treaty between the United States, Russia, England, and Japan put an end to wasteful open-water commercial hunting (in which half the animals sank out of sight when shot), and regulated commercial hunting on land. The population rebounded, reaching an all-time high in the 1950s. Then managers mistakenly allowed the commercial hunting of females, which led to a second slide in population. Though the practice was discontinued in the early 1960s, fur seal numbers continued to decline for another two decades. The Fur Seal Act was passed in 1966, and the population stabilized in the mid-1980s but has never quite recovered. A combination of disease, competition for prey from human fishing, and entanglement in drift nets cast off from commercial fishing vessels at sea may be responsible.

Fur seals breed primarily on two islands in the Pribilofs; a much smaller population has colonized San Miguel Island in California's Channel Islands. We should count ourselves lucky that fur seals continue to migrate and feed along the Northwest coast today, passing close to Cape Flattery and Cape Alava as they have for centuries.

Sea lions are regular visitors to the Olympic coast. They haul out in large numbers on offshore rocks, and their large, dark shapes are sometimes seen from the shore. Tatoosh Island off Cape Flattery, the Bodelteh Islands off Cape Alava, and Carroll Island (visible only by boat) are all favored haul-outs. Two species forage and rest together here. The northern or Steller's sea lion *(Eumetopias jubatus)* is the larger of the two, with males weighing about 2,000 pounds. Steller's sea lions are tawny in color, and adult males wear shaggy blond manes over broad necks and shoulders. The California sea lion *(Zalophus californianus)* is smaller; males are about half the size of Steller's, with dark brown to black pelage. Adult male California sea lions also have prominent crests on their foreheads. Both species possess large, powerful hind flippers that can rotate forward. These and their well-developed front flippers allow them to walk and climb around comfortably on land. As a result, sea lions can climb much higher on rocks than seals can. There, they sit up on their flippers or lounge about in the sun blissfully regardless of tides. Though it may be difficult to tell the two species apart without binoculars, one identification is nearly foolproof: California sea lions often bark like circus seals (most performing "seals" are actually sea lions), while their northern cousins remain regally quiet. Both differ from seals by their curled external ears, long front flippers, and their ability to walk on "all fours." A distinct population of Steller's sea lions numbering between 10,000 and 20,000 breeds from Alaska's Kenai Peninsula out into the Aleutian Islands. A southern population of about 20,000 breeds from southeast Alaska as far south as Santa Cruz, California. The male and female Steller's sea lions that visit the coastal and inside waters of the Olympic Peninsula disperse from these southern breeding territories, though some nonbreeders reside here year-round.

California sea lions breed in May and June, mostly among California's Channel Islands but as far south as Baja. Neither California nor Steller's sea lions breed along the Washington coast. This may be due in part to the steep and exposed nature of this coast's rocks and islands. Many haul-outs here are intertidal, and larger rocks and islands have steep slopes. Few are protected from the 15- to 20-foot waves that often pound the coast during breeding season. Large numbers of both sea lion species arrive to feed here in late summer and fall, however, and many stay through spring. From 4,000 to 6,000 California sea lions pass along the Olympic coast in late August and September en route to foraging areas in the Strait of Juan de Fuca and the inland waters of British Columbia and Puget Sound. Another 400 or so remain on the outer coast. The coastal population swells again in May as sea lions wintering in inland waters head back to their southern breeding grounds.

Steller's and California sea lions share haul-out and feeding sites rather amiably on the coast. During peak migrations in fall and spring, their numbers are about evenly divided; during the rest of the year Steller's hold sway. Codfish, smelt, squid, rockfish — almost any available middle and bottom fish — serve as prey for sea lions. Steller's take mostly pollock in Alaska; Californians concentrate on hake.

California sea lion numbers have been increasing steadily in recent years, and large numbers of males range north as far as British Columbia (only about six females have been identified in this area over the past 20 years). A small band of California sea lions has become infamous in Puget Sound by persistently feeding on a recovering steelhead run that returns through Ballard locks to Seattle's Lake Washington to spawn. Efforts by state and federal agencies to harass the feeding sea lions away from the site have been remarkably unsuccessful. Even capturing and removing the culprits to California has proved futile; the same individu-

Olympic National Park

als return to the Ballard locks in little more than a month. "Hershel," as the group has collectively become known, has made few friends among sport fishermen.

Obviously, California sea lions are doing well; Steller's are less fortunate. Although the southern population (the one that visits peninsula waters) is fairly stable, the northern population has declined precipitously. Their numbers have dropped from an estimated 100,000 in the mid-1980s to 10,000–20,000 a decade later, and more than 90 percent of their rookeries have been abandoned. This dramatic decrease caused this population to be listed as threatened under the Endangered Species Act in 1993. If the trend continues, they may become classified as endangered.

California sea lions migrate north in late summer and autumn to feed in Olympic's coastal waters and haul out on offshore rocks (sea lions are excellent climbers). Most return to southern breeding grounds in spring.

The problem may have to do with the animals' food base. Natural environmental changes, possibly triggered by El Niño–related events, may be a factor, but more likely are human impacts on their food sources. Pollock, their major food, is fished heavily by huge factory processors operating in the Bering Sea. Entanglement in fishing nets and illegal shooting may also be compounding the problem. Interestingly, the southern Steller's sea lion population — indistinguishable from the northern population and sharing the same natural history, except for a dependence on pollock — is doing fine, even increasing in numbers slightly. Research on this problem continues in northern waters. Meanwhile, another threatened marine mammal, the sea otter, has become a success story on the Olympic coast.

The Sea Otter's Return

A highlight of any trip to the outer coast is to look down from a headland or bluff and spot a sea otter playing among the kelp. Small, animated, and incredibly agile, sea otters *(Enhydra lutris)* differ from other marine mammals in that they lack a thick layer of blubber. Instead, a luxuriant coat of fur keeps them well insulated. So well, in fact, that sea otters seem loath to haul out on anything but cool, wet, kelp-covered rocks. As a result, sea otters along the Olympic coast haul out only at the lowest tides when suitable rocks are exposed. Sea otter populations to the north are much less discriminating about their haulouts, while otters in California rarely haul out at all.

Sea otters here are often confused with river otters. The latter are smaller animals that inhabit streams and coastal areas all around the peninsula. River otters forage along intertidal areas close to shore and are by far the most common otter seen on the outer coast. They are small (three to four feet, around 30 pounds) and weasel-shaped with brown faces and long slender tails. Sea otters are larger (up to five feet and 100 pounds). They are also brown but have grayish heads and faces, and short stout tails. River otters spend much of their time out of the water and never swim or float on their backs; sea otters rarely leave the water and spend most of their time on their backs, feeding and resting. They float high in the water, buoyed up by air trapped in their fur, which they aerate and groom incessantly. Their grooming rituals are a delight to watch. The animals twist, turn, and tumble in the water, scouring and puffing up every part of themselves, making their sleek, wet fur glisten.

Sea otters have a special relationship with the kelp forests that grow off Olympic's rocky shores — in fact the two are almost as inseparable as spotted owls and old-growth forests. Along with clams, crabs, and snails, one of the sea otter's favorite foods is sea urchins — and urchins feed on kelp. Nibbling at the stipes and holdfasts of giant and perennial kelp, sea urchins left unchecked can do serious damage to kelp forests, even to the point of eliminating them. When otters were extirpated from most of the California coast by the turn of the century, many of the kelp forests soon disappeared. Research conducted in the Aleutian Islands, southeast Alaska, and Vancouver Island has shown that predation by sea otters dramatically reduces the abundance of sea urchins, which in turn favors the development of kelp forests. These dense, tangly habitats attract an abundance of marine organisms, fish, and the birds and mammals that feed on them.

Though they feed both in perennial and bull kelp, sea otters here are particularly fond of perennial kelp *(Macrocystis integrifolia),* which grows in the more protected areas of the outer coast. This preference may have to do with the year-round protection and stability perennial kelp forests offer from waves. When the first storms hit the coast in fall, neighboring bull kelp — an annual — ends up in heaps on the beach. But otters and their pups wrap their webbed hind feet in perennial kelp and ride out swells and currents. Large numbers of otters are often seen rafted together in kelp beds off Cape Johnson and Makah Bay. Otters also seem to favor perennial kelp for diving, feeding, and pupping. Sea otters (and kelp) congregate north of Destruction Island, around Teahwhit Head, Cape Johnson, Jagged Island, Yellow Banks, Sand Point, and Cape Alava. I have recently begun to notice them off Cape Flattery as well.

Sea otters forage early in the morning and again later in the afternoon, resting, grooming, and playing in midday (the best time to watch for them) and rafting up toward evening. They are voracious feeders, consuming up to 40 percent of their body weight each day. Since shellfish are a major part of their diet, sea otters have developed the ingenious habit of cradling a flat rock on their chests as they float and using it to break open shellfish too large to crack with their teeth. Captive otters are known to keep their favorite rocks, tucking them under their arms as they dive. In California, otters' taste for abalone and crab has drawn the ire of some commercial fishermen, but biologists point out that many of these species reached artificially high numbers in the absence of sea otters, and sustainable harvest levels must be recalculated with the otters' recent recovery.

Coveted by humans for its beauty as well as its warmth, the sea otter's luxuriant fur was almost the animal's undoing. In the later 18th and 19th centuries, hundreds of thousands of sea otters were slaughtered by Russian and American fur traders. Hunts were unregulated, and traders moved quickly from one area to the next as populations disappeared. The last sea otter was shot on the Washington coast in 1910; the last professional sea otter hunters in Washington had quit the trade a few years earlier. By the time an international treaty was signed, too late, in 1911, sea otters were extirpated from most of their range, from Baja to the northern islands of Japan. Fortunately, remnant populations survived in the Aleutian Islands in Alaska, and in 1938, a small band was discovered off the Big Sur coast of California.

Under the treaty's protection, surviving populations began to recover. In the late 1960s and early 1970s, a concerted effort was made by wildlife agencies to restore sea otters to their former ranges in southeast Alaska, British Columbia, Washington, and Oregon. Fifty-nine sea otters from a healthy population on Amchitca Island in the Aleutians were introduced to the Olympic coast in 1969 and 1970. The animals released at Point Grenville in 1969 were not given an opportunity to acclimate to their new conditions, and most are believed to have died. But the 30 animals released at James Island the next year were allowed to acclimate in floating pens, and the introduction proved successful. Seven years later, 19 otters were counted; by 1981, the number had increased to 36. The population continued to grow, reaching the biological equivalent of "critical mass" in the late 1980s; then numbers shot up. A 1994 survey placed the population for the Olympic coast at 360. Because of their limited range and vulnerability to oil spills, Olympic's sea otters have been classified as endangered by the State of Washington. But since they came from a healthy parent stock in Alaska, they have avoided federal listing.

Ironically, just as the sea otter's uniquely rich fur nearly brought it to its demise a century ago, its fur puts it most at risk today. When coated or even spotted with oil, a sea otter's dense fur loses its remarkable insulating properties, and cold seawater soaks through the coat. With no blubber layer to act as backup insulation, the otter quickly becomes hypothermic and dies. Otters also sicken by ingesting oil while trying to "groom" it out of their fur. Today, the greatest threat to the interconnected community that exists on the outer coast — sea mammals, birds, and intertidal species — is the singular risk of spilled oil.

Oil and Water

Though the long progression of shipwrecks noted earlier in this chapter were fatal to most of the ships' crews, they were fairly benign to coastal marine life. Sailing ships had no need of fuel, and early steamers carried little that would damage fragile coastal environments. The period of grace ended in 1972, when the *General M. C. Meigs* ran aground in a winter storm and spilled 55,000 barrels of fuel oil onto the rocky northern coast. The spill did tremendous damage to birds and intertidal organisms. It also alerted managers of natural resource agencies to the dangers posed by ships and tankers operating off this rugged coast.

But it wasn't until December of 1985, when the single-hulled oil tanker *Arco Anchorage* ran aground in Port Angeles harbor and spilled 239,000 gallons of crude oil into the Strait of Juan de Fuca, that national park and national wildlife refuge managers snapped to attention. While it occurred well outside the boundaries of Olympic National Park, the grounding was just 10 miles east of Dungeness National Wildlife Refuge, home to some 15,000 wintering waterfowl. Even though the ship struck in a harbor adjacent to a Coast Guard station, it was more than three hours before containment booms were available to surround the spilled oil, and many thousands of birds became coated as the oil spread on the incoming tide. The vast majority sank beneath the sheen, but thousands were picked up along area beaches and taken to an improvised cleaning and rehabilitation facility in Port Angeles.

I remember Christmas of that year well, as my wife and I joined thousands of volunteers streaming into Port Angeles to help clean and care for the oiled seabirds. Washing the sick and listless grebes and murres made vivid the cost of our national addiction to oil and the risk that shipping oil poses to this fragile coastal ecosystem. Of the thousands of injured birds collected and treated (a small percentage of those affected), less than 10 percent survived.

Interagency coordination in dealing with the *Arco Anchorage* spill was virtually nonexistent. A few years later, when a lumber barge broke up off the outer coast and millions of board feet of lumber washed up on park beaches, biologists breathed a worried sigh. What if the barge had been one of the countless oil tankers plying coastal waters?

In December of 1988, their fear became a reality. The barge *Nestucca* was rammed by its own tug after it had broken free in rough waters just south of the Olympic coast. Nearly a quarter-million gallons of fuel oil spilled into the open ocean. Within days, tar balls and sheets of oil began to wash ashore. Then came the dead and dying birds: murres, auklets, scoters, grebes. Again, a rescue effort was mounted, but many birds were missed on the winter shoreline, and oil reentered the food web through scavengers. Of the 10,000 birds collected on the peninsula, more than 7,000 of them were dead; just over 1,000 were rehabilitated. The total estimated number of birds killed exceeded 55,000. Common murres suffered the greatest losses, making up 72 percent of all birds affected — 11 percent of the total murre population for the area.

Because most of the oil from the *Nestucca* came ashore in stormy conditions at high tide, it was the upper intertidal areas, driftwood logs, and kelp mats on the beaches that received the brunt of the oil. By sheer luck, the spill did not seriously affect the park's intertidal communities. Even so, coordination of cleanup efforts through a maze of state and federal agencies resulted in confusion, frustration, and crossed efforts. This led Olympic National Park officials to insist that the U.S. Coast Guard adopt a system to coordinate multiagency response to future spills similar to the one used by the Forest Service to respond to large wildfires in California that swept across jurisdictional boundaries.

The effectiveness of the system was tested three years later when another spill hit the Olympic coast. The Japanese fish-processing ship *Tanya Maru* was struck midship by a Chinese freighter 25 miles off Cape Flattery. The processor sank in a matter of minutes, releasing more than 200,000 gallons of diesel and other fuel oils into south-flowing currents that carried it onto the Makah Indian Reservation and the north Olympic coast. Again, the oil laid waste to seabirds: common murres, murrelets, auklets, and puffins. This time, the cleanup effort was well coordinated: oil-absorbent pom-poms were strung across beaches on incoming tides; rocks were scrubbed, and oiled driftwood logs and seaweed mats were removed by helicopter to keep oil from reentering the coastal ecosystem. Still, it became painfully clear that even with extensive planning and expert coordination, as soon as oil enters water, much of the battle is lost.

Although ships and tankers continue to pose the greatest immediate threat to the coastal resources of the Olympic Peninsula, another threat

Janis Burger

began to surface in the 1980s. The same year as the *Nestucca* spill, the U.S. Department of the Interior's Minerals Management Service under the Reagan administration proposed oil and gas leasing on the continental shelf off the Washington coast. The proposal brought strong opposition from tribes, fisherman, coastal residents, and conservationists nationwide. Their arguments were underscored a few months after the *Nestucca* spill when the *Exxon Valdez* released nearly 11 million gallons of crude oil into the pristine waters of Alaska's Prince William Sound. Clearly, greater protection was needed for

An oil-killed scoter washed up on the beach after the Nestucca *spill in December of 1988. More than 55,000 birds were killed by that single spill. Offshore oil spills continue to pose the greatest threat to the coastal ecosystem.*

the Olympic coast and its offshore waters. Responding to pressure from constituents in Washington and across the country, Congress directed the National Oceanic and Atmospheric Administration that year to designate a national marine sanctuary along the Olympic coast.

The Olympic Coast National Marine Sanctuary was dedicated in 1994. Congress had designated an area from Cape Flattery south to Copalis Rock and 30 to 40 miles out to the break of the continental shelf. Offshore oil and gas drilling is prohibited within the sanctuary, and vessels carrying hazardous cargo are requested to avoid sensitive areas adjacent to the coast. As the management plan for the sanctuary was being developed, the Department of the Interior rescinded the Navy's permit for practice bombing of Sea Lion Rock just south of Olympic National Park, restoring peace to nearby nesting and haul-out sites.

By extending much-needed protection to offshore and nearshore marine areas, the Olympic National Marine Sanctuary is an important addition to the Washington Islands National Wildlife Refuges and the magnificent coastal wilderness of Olympic National Park. These agencies, which represent some of our government's best aspirations toward the natural world, join Native Americans of the Makah, Quileute, Hoh, and Quinault tribes in a long tradition of stewardship for this splendid joining of land and sea.

9
Footprints on the Land

Afternoon clouds were just settling over the mountains and foothills above the Dungeness River, but the valley fields were still bright with spring sunlight. The high shoulders of Blue Mountain and Deer Ridge were edged in winter snow, and the blue silhouette of Mount Angeles rose abruptly over the green slopes of Lost Mountain to the west. I crossed a field of wet spring grass where the foothills met the valley floor and stopped by a small marsh. Dry cattails rustled in the cool north wind, and a pair of mallards dabbled in the shallow waters of a winter pond.

It wasn't hard to imagine this low, rolling landscape in late Pleistocene time. The same ice-rounded foothills and hummocky piedmont plain stretched north to the Strait of Juan de Fuca. Just as now, the snowpack would be melting back from the mountains. The small marsh and pond was larger then, and bordered then as now by a thick growth of cattails. Instead of the grassy fields of the valley and second-growth Douglas-firs and cedars covering the foothills, there were shrubby patches of willow and alder interspersed with sedges, grasses, trailing blackberry, and wild rose. And in place of the houses lately sprung up in the old farmlands of the Dungeness Valley were slowly melting blocks of ice, stranded remnants of the recent advance of the Cordilleran ice sheet that pressed against the northern Olympics like a wedge.

This was the landscape the earliest known inhabitants of the Olympic Peninsula encountered more than 12,000 years ago. We know this because it was here in the late 1970s that a valley resident turned up one of the most significant archaeological finds in the Pacific Northwest. Emanuel Manis was deepening a pond for wintering waterfowl when he uncovered what he recognized as the tusks of a prehistoric mammal. Archaeologists from Washington State University examined the site and confirmed the animal was a mastodon, a large elephantlike mammal that inhabited Pleistocene landscapes. Within hours of beginning work, one of the archaeologists made a remarkable discovery. Embedded in one of the mammal's ribs was a broken piece of antler or bone resembling a spear point. The fragment was lodged more than three quarters of an inch deep, suggesting a blow delivered with great force. Further examination under a microscope revealed that the outer covering of the fragment was worn away, typically the result of human workmanship. The rib had partly grown over the spear point; this was not a fatal wound.

Other evidence indicated human involvement as well. A number of bones exhibited telltale signs of butchering, and the skull, which had been turned 180 degrees out of its natural position, was smashed into thousands of pieces. Several bones from the upper side of the skeleton were moved a short distance away, presumably out of the pond in which the carcass lay; these also showed signs of butchering. The skeleton was buried in stream-deposited sand and gravel, which lay directly atop the rocky rubble of glacial drift. Seeds and bits of wood recovered from the strata that contained the bones were radiocarbon dated at more than 12,000 years. The discovery was not only the earliest evidence of human presence in Washington, but the earliest association of Paleolithic hunters with mastodons anywhere in North America.

Not far from the pond is a small grassy rise alongside the road. It was the only high ground close to the watering hole, and researchers thought it might be the hunters' campsite. A cross-section of the rise revealed the charred remains of six to seven fire pits, all in different-aged strata, as well as the bones of several bison, caribou, muskrats, and ducks, and a tooth from a very large black bear. There was evidence of eight or nine layers of occupation by hunting people at this site between 6,000 and 12,000 years ago.

Early prehistoric hunters and gatherers probably followed the large Pleistocene herbivores to the northern peninsula from the south after the ice had retreated and vegetation began to reclaim the scraped and rocky landscape. As the climate warmed and became drier, human populations in the Great Basin and Columbia Plateau gravitated to rivers and lakes. Population pressures might have prompted some groups to explore the recently deglaciated landscapes to the north. Their dwellings must have been temporary, sewn hides or woven mats over stick frames; none have survived the passage of time. The people carried a few simple tools of bone and stone (a flaked cobblestone and a flensing tool were found at the Manis site), but little else is known of them. Who were these earliest of wayfarers on the peninsula, these small nomadic bands who hunted the raw glacial landscapes for mastodon, caribou, bison? What brought them to this remote and rugged corner of land, and what was their relation to the people who followed them thousands of years later?

Archaeological investigations in Olympic National Park have been fairly limited in the past. The park's rugged topography and thick forests proved daunting, and the mistaken belief that Native Americans were loath to venture into the interior mountains dissuaded early investigators. But archaeologists have taken a closer look at the mountains and river valleys of the park in recent years, and the picture of early human presence here is becoming much more complete. Early hunters left traces in the Olympics that lead from this small marsh at the edge of the northern foothills, up through the mountains, and deep into the interior ranges. The cultural artifacts that reappear on the coast thousands of years later offer a stunningly clear window into the sophisticated ways of sea mammal hunters whose descendants still inhabit the area. Though much of the archaeological record is missing — only fragments and a few simple stone tools remain from great sweeps of millennia — the story of how the earliest inhabitants of the Peninsula learned to adapt and thrive within this dynamic ecosystem unfolds like a blossom in a newly greening landscape.

The Mountain Hunters

From the Manis site, Blue Mountain dominates the southwestern horizon. A gentle, broad-shouldered peak, with fingers of forest reaching nearly
to its summit, its south side opens into the subalpine meadows of Deer
Park, named for the creatures drawn to its warm, south-facing slopes in
spring. Deer Ridge swings east from Deer Park in a long green undulat-
ing line before plunging south into the valley of the Gray Wolf River.
The ridge drops abruptly at a place called Slab Camp where a tongue of
Cordilleran ice pushed through into the steep canyon of the Gray Wolf.
From the flat, glacier-carved gap that remains, it's half a day's walk to the
meadows of Deer Park, an even shorter walk down to the Gray Wolf.
Sheltered from prevailing storms and watered by a perennial stream, Slab
Camp was the site of an old Forest Service ranger station for years; today
it hosts a primitive campground popular with hunters in deer season.

Archaeological work done at Slab Camp in the 1980s uncovered
extensive evidence of early human use of the area. The site yielded no
hearths or animal remains, but more than 64 stone tools were discovered
— points, flakes, hammer stones, and cobble tools — as well as thousands of worked chips. The artifacts were located below a layer of
volcanic ash from Mount Mazama, a volcano that erupted in the southern Cascades about 6,800 years ago (the clear waters of Oregon's Crater
Lake now rest in its caldera). Mazama's ash spread north and east as far
as British Columbia and Montana, and the ash and pumice deposit it
left behind provides a handy benchmark for archaeologists. This date,
and the nature of the artifacts recovered, show the Slab Camp occupants
practiced a way of life archaeologists describe as the "coastal Olcott pattern," named for the family farm in the Stillaguamish River east of Puget
Sound, where the pattern was first identified.

Olcott people were "generalists"; they hunted land animals, gathered
plants and berries, probably fished. Little is known of these early inhabitants, other than the scant evidence gleaned from the stone tools they
left behind. They chipped basalt into simple, expedient tools; many of
these tools appear to be quickly made, possibly on site. They also fashioned more elaborate blades and spear points of varying sizes. These
were worked into the distinctive shape of a willow leaf, an indicator of
the coastal Olcott pattern.

Creators of the Olcott pattern practiced a widespread subsistence lifestyle on the Northwest coast that remained stable for several thousand years, roughly from about 9,000 to 6,000 years ago. Their way of life is well represented on the peninsula. Olcott sites near Quilcene, Port Angeles, and at Lake Cushman in the southeastern Olympics yielded distinctive "willow leaf" spear points, and an Olcott point was found just above Mazama ash at the Manis site. A number of other Olcott-pattern tools turned up in these sites as well. Choppers and scrapers indicate that people were working animal hides; abrading stones suggest the manufacture of bone awls or needles, probably used for making clothing or basketry, and gravers point to working with wood or bone. This hunting and gathering way of life raises intriguing questions: Does this land-based pattern indicate that people had not yet learned to use nearby marine resources? Or were these hunting camps merely the inland component of a seasonal round that included coastal sites now lost to erosion and changing sea levels?

At this point no one can say. During much of Olcott-pattern life on the peninsula, sea levels were considerably lower than today. The rebounding of the land surface after the weight of the ice sheet was lifted and resulting fluctuations in sea levels would have placed most coastal sites of those days below current sea level. This, combined with the dense junglelike growth along the coast and the erosive power of the waves, may have conspired to hide or eliminate an important component of Olcott-pattern life here. But evidence of early human use of Olympic National Park's high country abounds.

From Slab Camp, the subalpine meadows of Deer Park are easily reached by following open south slopes along Deer Ridge. A large archaeological site at Deer Park yielded artifacts similar to those found at Slab Camp: points, flakes, a cobble tool, unfinished points and blades, an abrading slab. There was no organic material by which to date the site, but the tools match the Olcott pattern. Interestingly, the basalt from which they were fashioned seemed different from native Crescent basalt. In fact, the rock is dacite, a glassy, fine-grained, extrusive rock similar to andesite. Dacite weathers slate gray and fractures into smooth, shiny black faces; it feels lighter in the hand than Crescent basalt, and it flakes to a finely honed edge. Cobbles and boulders of dacite are believed to have been rafted to the Olympic Peninsula by the Cordilleran ice sheet from parent rock in the Howe Sound area of British Columbia. The rocks are still commonly found in streams and along shorelines that lay in the path of the continental ice sheet, largely below 3,500 feet. The presence of dacite in the interior mountains or above the height of the last ice sheet means that humans brought it there.

If you trace the interconnecting ridges on a topographic map from Deer Park and Blue Mountain west to Obstruction Point, then farther on to Hurricane Ridge or south around the Gray Wolf headwaters, you'll notice vast stretches of open subalpine country joined by ridges and watered by lakes and snowmelt streams. The high country here is mantled with lush meadows, pocketed with marmot burrows, and criss-crossed with the trails of deer and elk. Nearly everywhere they have looked, archaeologists have found ample evidence of human use of these areas in the form of dacite flakes, points, and worked cobbles. Recent investigations by park archaeologists have shown that early hunters and gatherers left traces even farther into the interior, along the open ridges of High Divide and the remote Bailey Range. The places where they camped, perched on a scenic overlook, or stopped to rest in the shade of a rock outcrop are often the same places backcountry hikers stop and rest today.

I find a deep satisfaction in thinking of the old ones — "the People," as early Native Americans everywhere referred to themselves — ranging over these mountains with a few simple, elegantly honed tools. I think they must have shared something of the delight we feel in the rugged splendor of this mountain landscape. Their stone whittlings scattered across upland ridges and basins are words spoken across a changing landscape of ten thousand years. They add a familiar element to the haunting beauty of these coastal mountains. And they cast our wilderness wanderings today in a different light.

Only one high-country site, by a tarn in Seven Lakes Basin in the north-central Olympics, has yielded a positive date for early human use of the high country. Charred ash from a fire pit buried beneath a foot of organic litter was dated at about 5,000 years before the present. Chipped dacite and charred bones suggest the site was used as a hunting camp. By that time the period of postglacial warming had ended and the climate had cooled in the Olympics. Sea level was beginning to stabilize, and vegetation had come to resemble that of today. The elk and deer who summer in this moist, north-facing basin today may be descendants of the animals that drew small hunting bands from nearby lowlands in summer and fall. Perhaps those distant travelers were gathering meat to dry for winter use, or simply enjoying the largess of a fall high-country hunt. Either way, the Seven Lakes site suggests a seasonal use of high-country resources at a time when seasonal adaptations to marine resources were also beginning to appear on the Northwest coast.

Early Coastal Cultures

To trace the path of early maritime culture in the Pacific Northwest — from hunting of land mammals to use of nearshore resources — it's necessary to venture slightly north, to the Fraser River Gorge in mainland British Columbia. There, a site discovered on the grounds of an old cannery shows evidence of stable, long-term human occupation from 8,000 to 2,000 years ago. Stone hunting artifacts of the Olcott pattern dominate the lower archaeological strata, but about 4,000 years ago, increasing numbers of bone and antler tools began to appear. Chopping cobbles became rare around that time and abrading stones for wood and bone work much more common. Chipped stone points became smaller, and ground slate points appeared. New tools also made their first appearance: carving tools made with rodent teeth for detailed woodworking, mussel shell adze blades, and antler wedges.

But the most significant difference from earlier times was a marked increase in the use of marine resources. Remains of salmon, sturgeon, and other fish as well as mussels and seabirds were added to the earlier remains of land and sea mammals. Antler wedges suggest large-scale woodworking, and it is very likely that seagoing canoes had been developed by this time. Shellfish became a winter staple, marking the beginnings of a seasonal adaptation that would soon unfold into the later maritime and Northwest Coast culture.

A similar evolution of hunting and gathering strategies was doubtless unfolding on the Olympic Peninsula as well — particularly as the climate cooled and forests began to close. The first evidence of this early maritime way of life on the peninsula was discovered at the mouth of the Hoko River, 15 miles east of Cape Flattery on the Strait of Juan de Fuca.

The Hoko River drains wet coastal lowlands in the northwest corner of the peninsula. Running brown and muddy during winter storms, the river deposits tons of silt along its lower floodplain and on its delta on the Strait of Juan de Fuca. Here, a rich assemblage of wooden artifacts preserved in the wet mud of the river mouth shows that the site has served as an important seasonal fishing camp for nearly 3,000 years. Radiocarbon dates from Hoko indicate regular occupation of the site from 2,200 to 2,900 years ago. The mouth of the Hoko remains a favorite fishing spot for Makah people in whose traditional territory it lies.

Sealed from oxygen and organic decay in the wet mud of the river mouth, well-preserved wood and fiber artifacts and abundant animal remains paint a vivid picture of an early coastal and riverine fishing culture. By carefully excavating artifacts with spray from hoses pumped from the river, archaeologists uncovered bent-wood halibut hooks used in offshore fishing, bone or antler harpoon heads for hunting sea mammals, a fragment of a weighted split-spruce fishing net, and a possible fish weir. Also washed from the mud were wooden wedges, carved points, and a piece of adze-hewn cedar board.

One fascinating find at Hoko included six intact knives made by lashing flakes broken from quartzite beach pebbles into split-cedar handles. Replicas of these knives proved excellent for cleaning and filleting fish. The number of knives and blades recovered from the site, along with fragments of burden baskets, suggest that the people who camped here processed large numbers of fish, most likely for storage and later use. The hooks and net fragments indicate they fished the offshore bank, and remains of fish from offshore waters were twice as plentiful as salmon, which were most likely speared in the river. Canoes were probably used to take so many offshore fish, though no canoe fragments were found at the site. Nor was evidence of plank houses, though tools for splitting large cedar boards found elsewhere make archaeologists suspect that such dwellings were part of early maritime culture. Remains of seals, porpoises, and dolphins show that these sea mammals were hunted regularly, probably using the barbed spear points found at the site. Though the bones of a gray whale were found, no whaling tools accompanied them, suggesting the whale was scavenged from the beach.

Olympic National Park

A presage of the flowering of the Northwest Coast culture to come was also washed from the mud of the Hoko
River. A redcedar tool identified as a mat creaser, used to
soften cattail or rush fibers to be woven into mats, was
carved into the likeness of a pair of kingfishers joined at the
beak tips. This stylized design, simple and elegant, is the
oldest artwork found anywhere on the Northwest coast.

A Makah fishing camp on Tatoosh Island, showing cedar canoes and halibut drying racks. The photograph was taken by Samuel Morse, an agent at Neah Bay from 1890 to 1903.

At least three other early maritime sites have been discovered on the peninsula, two on the outer coast and one on the strait.
All are shell middens containing bone artifacts and antler woodworking
tools, along with plentiful remains of sea mammals and shellfish. Their
location on uplands and terraces above today's coastline points to a
higher sea level at that time. Like the lower strata at the cannery site in
British Columbia, test excavation of one of these sites shows widespread
use of worked stone tools at lower levels. But this earliest of technologies, by which humans lived here for the last 12,000 years, all but
disappears from northern coastal sites on the peninsula about 1,000
years ago. Instead, Native Americans developed highly refined skills for
detailed working of wood, shell, and bone. This mastery of the
resources of forest and sea led to the development of the elaborate cultures that followed.

Ozette: Life in an Early Whaling Village

Contact with European seafarers in the later part of the 18th century forever changed the lives of the indigenous people of the Northwest coast. New technologies and a different kind of material wealth introduced by fur traders disrupted long-standing social traditions, alliances, and mores. The introduction of exotic diseases wiped out entire villages and devastated Native American society. Treaty policies of the 1850s and the government schools that followed often outlawed Indian cultural practices and beliefs, and generations of Native people became exiled from their past.

Early observers writing in the wake of these social upheavals often reflect a skewed and somewhat limited view of indigenous life on the coast. Many details of the ways in which Native Americans interacted with each other and with the ecosystem were lost. Archaeological knowledge of early Northwest lifeways was also limited by the perishable nature of the wood, bone, and fiber materials that supported coastal culture over the past 1,000 years. To see how native cultures here evolved with the natural systems of the Olympic Peninsula, it's necessary to look to the time before the first contact with Europeans. Fortunately, on the northern end of the Olympic coast, one of the most important archaeological sites in North America reaches back to that time and offers a detailed look at daily life in a prosperous village of sea mammal hunters.

Ozette Village is situated in the lee of Cape Alava, just south of the Makah Indian Reservation. A photograph taken of the village in the late 1800s shows a cluster of a dozen or so houses. Most are built of split cedar planks, and a few are in the old shed roof longhouse style. One of five permanent Makah villages located around Cape Flattery on the northwest tip of the peninsula, Ozette remained occupied well into the 20th century. Families leaving the village to be with children attending mandatory boarding school in Neah Bay eventually led to Ozette's demise. Some returned to hunt and fish in summer, but by the 1930s, most of the houses had tumbled down and the hillside was overgrown with trees and brush.

Ozette contained the most extensive undisturbed shell midden ever found on the Olympic coast, however, and it was known to have been a principal whaling village for the Makah. In fact, its location was ideal for sea mammal hunting. Situated on the westernmost point of land in the contiguous United States, the village lay close to the migratory routes of whales, fur seals, and other sea mammals. Large offshore reefs and islands offered protection from pounding Pacific swells and provided easy passage for canoes.

A fairly detailed picture of Makah whaling culture has been pieced together from ethnographic reports and accounts from tribal elders. According to tribal traditions, whale hunting defined Makah society. The endeavor was undertaken only by whaling chiefs, an inherited privilege, and with a great deal of prescribed ritual and religious observance. Pursuing gray or humpback whales at sea — eight men to a 35-foot cedar whaling canoe, armed with harpoons tipped with mussel shells and seal-skin floats — was a difficult and dangerous undertaking. Whalers fasted, purified themselves, and prayed to their spirit guides well in advance of the spring hunts, and they and their family members adhered to strict taboos. A whale could sustain a whole village for weeks with plenty of oil and blubber left for trade.

Whalers were afforded tremendous prestige, and whaling families formed an aristocracy. Like most indigenous societies on the Northwest coast, Makah society was stratified. The noble class consisted mostly of whaling chiefs

Ozette Village at the end of the 19th century. The remains of the ship Austria, *shipwrecked in 1887, litter the beach in the foreground.*

Clallam County Historical Society

and their families; successful hunters enjoyed a high degree of wealth and status. The broad class of commoners possessed a variety of learned skills, from basketry and canoe carving to seal hunting. The underclass of slaves (probably about five percent of the population at time of European contact) were captured in battle, traded from other villages, or born into slavery. By and large, they led lives of servitude.

Nobles enjoyed privileged access to prime hunting, fishing, and gathering sites throughout the Northwest coast; these rights were inherited or acquired through marriage. Members of noble families were expected to marry within their class, among different villages or tribal groups. Despite this strict social order, all classes enjoyed a good deal of social mobility. Commoners could better their standing and increase their wealth through marriage, by acquiring a specialized skill, or through warfare. They could also shift allegiances among chiefs. All classes — nobles, commoners, and slaves — lived together in the longhouses at Ozette.

In the 1960s a team led by Washington State University archaeologist Richard Daugherty excavated a trench in the large shell midden at Ozette. Working from the beach up through successive layers of coastal terrace, their shovels turned up more than 2,000 objects. Some — buttons, coins, rifle parts — were obviously from the historic period. But layers beneath them predated European contact. Bone fish hooks, stone sinkers, bone and shell blades, and carved combs dated back an estimated 2,000 years. The abundance of fish and sea mammal bones throughout the midden showed that a stable and highly organized hunting and fishing society had inhabited the site almost continually during that time. Nearly all of the bones found were from sea mammals, predominantly fur seals, but the discovery of whale bones confirmed what had long been known by Makahs; Ozette was indeed a whaling village, possibly one of the largest south of Alaska.

In the second year of initial excavation at Ozette, archaeologists sunk a test pit in a wet area back from the beach. Deep in the mud, they unearthed a coiled rope made of cedar boughs. Further excavation revealed fragments of woven cedar bark mats and baskets, along with a section of hewn cedar house plank — all perfectly preserved in the wet clay. Makah oral traditions told of landslides that buried Ozette Village in the distant past; here was evidence. Like the alluvial deposits that preserved highly perishable artifacts at Hoko, clay had apparently preserved similarly perishable artifacts at Ozette. If the oral traditions were accurate, here, beneath the shrubs and young trees, might lie a virtually intact whaling village — complete with wood, fiber, and bone artifacts — sealed since the day it was buried. A find of this magnitude required major excavation, and Daugherty and the Makah Tribal Council began planning one. As it turned out, the Pacific Ocean had plans of its own.

A severe storm early in 1970 battered the upper beach with waves and began to erode the sea bank in which the collapsed longhouses were buried. The midden slumped, and house planks and artifacts were exposed: an inlaid box, a canoe paddle, a woven rain hat. Daugherty and the Makah Tribal Council secured emergency funding so that excavation could begin that spring. Forgoing shovels and trowels for seawater pumps and hoses, spray nozzles and fine-bristled brushes, teams of archaeologists and students from around the country began the work of unearthing the first houses. Artifacts sealed in clay for 450 years were coaxed once more into coastal sunlight: delicately wrought combs, whale bone clubs used for seal hunting, dozens of finely woven baskets, and elaborately carved wooden bowls. Not only finished pieces but artifacts in various stages of manufacture and repair were also recovered, offering a rare cross section of a functioning coastal culture. Piece by carefully excavated piece, the story of early everyday life on this coast — and something of the people who lived it — was revealed.

A burden basket was washed free of the clay, filled with rolls of cedar bark newly stripped from a tree and not yet processed into fibers for mats, baskets, or clothing. Spindles and pieces of looms emerged, along with an intricately woven plaid blanket in a pattern previously unknown on the coast. Other items emerged from the clay: carved bent-corner cedar boxes used for cooking with heated stones, larger decorated storage chests, rain hats, halibut hooks, wedges, and woodworking tools. Some chisels were fitted with steel blades, possibly salvaged from shipwrecked Japanese junks. Woodworking technologies evident in earlier archaeological finds had obviously been carried to a high level on the coast by this time. The level of detail and distinctive style of the decorative artwork suggested a highly developed esthetic.

The lay of the houses and artifacts recovered also gave clues to the structure of Makah society. Whaling equipment was often found at the rear of the houses, places occupied by whaling chiefs and their families.

Sleeping platforms covered with cedar mats lined the walls, and each household maintained its own hearth, sometimes separated from the others by screens.

The mudslide that buried — and so preserved — Ozette village came so suddenly that there was no time to retrieve valuables. Inside a cedar bark sheath lay a finely honed harpoon head of mussel shell with carved elk bone barbs and fitted with woven fiber rope (similar points were found broken off in whale bones). A long cedar plank was carved with the likeness of a whale that matched the petroglyphs at "Wedding Rocks" a short distance down the coast. Wood and bone handles of everyday tools were carved in the stylized likenesses of people, owls, wolves.

"Wedding Rocks." It is impossible to date these petroglyphs south of Cape Alava on the coast, but their style closely matches designs carved into artifacts dated at close to 500 years old found at Ozette Village.

Janis Burger

Some objects seem to have been made for ceremonial or purely artistic purposes. A bone carving of a small human figure crouched within a mussel shell echoes a Makah creation story; and a beautifully carved cedar whale fin was inlaid with 700 sea otter teeth in the motif of the mythical Thunderbird. The striking beauty of these designs reflect not only the Ozette people's reverence for the natural world that sustained them, but their close ties to a spirit realm that permeated that world and insured their well-being and prosperity.

The archaeological dig at Ozette lasted 11 years. Three of five known longhouses were fully excavated and over 50,000 artifacts recovered. All remained with the tribe. Many of the finest pieces from this collection can be seen at the Makah Cultural and Resource Center in Neah Bay. For the Makah people of today, Ozette is a gift from the past offering a deeper perspective on their own rich heritage, a fuller sense of who they are. For all of us, Ozette is a window on a stage of the human journey on this coast, from nomadic hunters following game across the spare, post-glacial landscape to sophisticated village dwellers living amid plenty and sharing such "contemporary" pursuits as trade, status, and the cultivation of artistic expression.

Northwest Coast Adaptations

The Makahs' way of life reflected their unique position at the peninsula's northwest tip. Access to migratory fur seals at Umatilla Reef and halibut banks off Cape Flattery as well as Ozette's proximity to the migratory routes of whales influenced many of their cultural practices. The location of other Native American villages at the mouths of peninsula salmon streams similarly influenced their cultures. The Makah shared many of their cultural practices with other Wakashan-speaking whaling people of Vancouver Island (the Nitinat and Nootka), but they also shared important cultural traits with their peninsula neighbors. All tribal groups on the peninsula shared a broad cultural pattern practiced from Yakutat Bay in Alaska to Cape Mendocino in California. This "Northwest Coast culture" is one of the most elaborate and sophisticated indigenous cultural patterns north of Mexico.

Generally speaking, Northwest Coast people are distinguished by an orientation toward marine resources, communal winter villages, extensive use of canoes for water transport, and a strong reliance on salmon. Northwest Coast cultures are also known for their specialized adaptations to particular environments and the skilled use of a number of widely scattered seasonal resources. Both traits are well illustrated among the tribal groups of the peninsula. Though peninsula peoples shared a number of traits, the wide cultural diversity found among these groups — their languages, beliefs, ceremonies, artistic expressions, and subsistence strategies — reflect the natural diversity of the Olympic ecosystem itself.

In winter, when heavy rains and long hours of darkness descended, the people of the peninsula gathered in communal winter villages. A village consisted of up to 20 longhouses, though most had only a few. The houses were usually about 30 feet wide and 60 feet long, constructed of split or adze-hewn redcedar planks lashed to post and beam frames with cedar withes; they were usually set along a river or shoreline. Two to six related families shared each house, each family maintaining its separate household. Wide cedar planks lined the walls, serving as beds and seating, and cedar baskets or bent-corner cedar boxes held family possessions.

Villages were occupied throughout the winter months, some year-round. In spring, summer, and fall, smaller family groups usually traveled to hunting, fishing, and gathering sites: shellfish beaches, salmon streams, inland lakes, prairies. Surpluses were gathered, processed, and stored at these seasonal settlements, supporting winter village life and affording the leisure time to develop the traditions of elaborate winter ceremonials that were a central part of Northwest Coast life.

"Tribes," as we've come to think of them today, were actually loose associations of villages at the time of European contact, sharing languages, kinship ties, and common food gathering locations. The formal tribal structure was imposed by the United States government with implementation of the treaties of the 1850s. Villages as well as "tribes" traded and intermarried, establishing important alliances and renewing intertribal relations. Central to this social order was the potlatch. Every momentous event in village life, every claim to hereditary privilege, whether a name or song or a prized fishing site passed on to a new generation, required a potlatch. Preparations for these elaborate gift-giving ceremonies sometimes took years. Guests traveled from distant villages, and the feasting, dancing, and singing that preceded gift-giving often went on for days. A village headman or host family gained great prestige by giving away amassed wealth. The songs and dances affirmed the heritage and position of the host family, and the value and order of gifts distributed acknowledged the status of the guests. By accepting the gifts, guests acknowledged the hereditary claims, prosperity, and social status of their hosts, thus reaffirming a family's place in the strict social order.

Potlatches also served another important function in Northwest society. Though the region is blessed with unusual abundance, it is also subject to unpredictable seasonal shortages. Individual salmon runs can fail because of flooding or drought; changes in weather patterns can cause shifts in sea mammal migrations; occasional "red tides" may contaminate shellfish; a berry or camas crop may fail. Along with its other functions, the institution of the potlatch served to redistribute resources and "share the wealth." Since well-to-do families were expected to reciprocate in kind, a mutual sharing was assured. In spite of this kind of cooperation, warfare among tribal groups and even among villages was not uncommon. Some crimes and social improprieties were deemed so egregious they could be resolved in no other way. Raiding was also an accepted way to obtain slaves or increase one's status.

Whether raiding or visiting, hunting or fishing, all transportation along the peninsula's coastlines, rivers, and lowland lakes was by carved western redcedar canoe. The technology of canoe manufacture, from felling the giant redcedars to carving or painting the bow designs, was attended by a great deal of ceremony. Cedars were felled using a combination of chopping and controlled burning, then were split and hollowed out with adzes and burning coals.

Most tribal groups are known to have employed at least six sizes of these versatile craft. Small "shovelnose" canoes were popular on inland rivers and along parts of Hood Canal; larger sealing and whaling canoes of the open coast were extremely seaworthy under a variety of conditions, and huge 60-foot seagoing canoes were used for long-distance transport. Canoes were also an important trade item among tribal groups, with most of the larger seagoing canoes originating on the northern coast.

One tree, the western redcedar, formed the foundation of Native American culture on the Northwest coast. Lightweight, decay-resistant, and easily split into boards and huge house planks, redcedar was both versatile and lasting. Houses, canoes, clothing; baskets, boxes, and cooking containers; masks, headdresses, and rattles; ropes, mats, and spear and arrow shafts; cradles, dolls, and even diapers — all were fashioned from its dark aromatic wood or reddish brown bark. Western redcedar and the salmon that spawned in its shade along the peninsula's rivers and streams were the heart and soul of Northwest culture. Human life here without them — whether in the distant past or the near future — is difficult to imagine.

Indigenous people on the peninsula were masters of woodworking and knew the distinctive qualities and appropriate uses of a wide variety of woods. Hard, resilient yew was best for bows and harpoon shafts; light, straight-grained cedar for boxes and boards. The hardwoods, maple and alder (which wouldn't flavor foods), were used for bowls; tough, flexible spruce root for burden baskets and rope; young Douglas-fir for dip net handles and spears.

Inhabitants of the Olympic Peninsula were incredibly imaginative and resourceful in making good use of the abundant resources that surrounded them. Foods were plentiful nearly everywhere on the peninsula in season, and tools made of wood, shell, and fibers allowed people to harvest and preserve it in quantities. Saltwater fish were taken with hooks and lines, gill nets, and seines; anadromous and freshwater fish with weirs, dip nets, and spears. Shellfish of many varieties were gathered in large numbers from intertidal areas all around the peninsula; surpluses were smoked, dried, and stored. During times of seasonal shortage of other foods, shellfish beds could prove sound village bank accounts. More than 50 species were identified in a single midden in La Push, and records show that large quantities of dried shellfish were sold to white settlers on the peninsula.

All tribal groups on the peninsula hunted land mammals as well as sea mammals. Elk and deer supplied meat, hides, antlers, and bone (land mammal bones are a much harder and denser medium for tools and weapons). For tribal groups that maintained upriver villages, like the Quinault, Quileute, Elwha S'Klallam, and Skokomish, land mammals may have formed an important part of their diet.

Berries were gathered throughout the summer and fall, from early ripening salmonberries along streams to late-blooming blueberries in the mountain meadows. Eaten fresh or dried (often mixed with seal fat), berries, along with bracken fern roots and camas bulbs gathered from open prairies, were among the principal plants used for food. Over 150 plants are known to have been used by Native Americans in western Washington, for foods, medicines, and materials.

Other Peninsula Cultures: The Quileute and the Hoh

South of the Makah villages on the Olympic coast lies the territory of the Quileute and Hoh people. Their traditional territory included the Quillayute River and its tributaries, the Dickey, Sol Duc, Calawah, Sitkum, and Bogachiel, as well as the Hoh River. The Quileute and Hoh are Chemakuan speakers, a language group unique to the Olympic Peninsula. At time of European contact they occupied three principal winter villages at the mouths of the Quillayute River (the site of the present Quileute Reservation at La Push), the Hoh River (site of the present Hoh Reservation), and Goodman Creek, as well as smaller settlements upriver and at other locations. They also maintained some 40 seasonal settlements scattered along the coast and inland river valleys.

For countless generations, Quileute people enjoyed what may have been the most varied way of life on the Peninsula. Though tremendous runs of salmon and steelhead filled the Quillayute and Hoh rivers, Quileutes also excelled at hunting sea mammals, taking seals and whales on occasion. They fished offshore and gathered shellfish, mounted elk hunts up the inland valleys, and gathered bracken fern roots and camas bulbs from Quileute and other prairies. Camas was an important trade item with the Quinault to the south, who traded sockeye salmon in return. Like many tribal groups on the peninsula, the Quileute kept their prairies productive and held back the encroaching forest by burning (today's town of Forks was settled on the Forks Prairie).

Like the Makah to the north and Quinault to the south, the Quileute fashioned highly decorative artwork. Quileute women were known for their watertight cooking baskets and for blankets woven from the hair of "wool dogs" kept for the purpose and from mountain goat wool imported from the mainland. Quileute art also reflected strong ties to the spirit world: to *Kwati*, the Changer, who created the first people from wolves, and to *Hayak*, the Raven who stole back the sun from the pow-

University of Washington Libraries

A native whaler prepares to harpoon a surfacing gray whale. Sealskin floats attached to the harpoon head will drag as the whale tries to dive, eventually exhausting it. (Photo: A. Curtis)

ers of darkness. The spirit world was also celebrated by five secret societies, which initiated participants into the ritual lore of the tribe. Initiates into the "wolf society" gained warrior spirit power; another society harbored fishermen. One small secret society was the exclusive domain of whale hunters; members of the "weathermen society" were said to exercise power over weather and the migrations of whales; and elk hunters followed the dictates of the hunters' society. Only the latter is believed to be entirely Quileute in origin; the other societies had counterparts among the Quinault and Makah. Unfortunately, little remains of the pre-contact religious life of the Quileute. A fire set by a white claimant to tribal land in 1882 destroyed all 26 houses at La Push, along with every tool, mask, basket, and hunting or ceremonial artifact.

Contact came early for the Quileute, and they apparently wanted no part of it. In 1775, the Spanish schooner *Sonora* lost a landing party and its longboat to Quileute warriors, as did the British ship *Imperial Eagle* a dozen years later. Shipwrecked crew members of the Russian-American Company ship *Sviatoi Nikolai* were captured as slaves in 1808, though most were later ransomed. Quileute warriors had a reputation for fierceness, and they conducted raiding and retribution parties along the coast from Vancouver Island to the mouth of the Columbia. James Island just off La Push, a burial place for Quileute chiefs, also served as a nearly impenetrable fortress while the village was under attack.

Besides the Quileute and the Hoh, the only other members of the Chemakuan language group were the Chemakum people of the Port Townsend area at the northeast corner of the peninsula. Quileute legend tells of a great flood that separated them from the Chemakum. The Chemakum people were a small, isolated band, fending off hostile neighbors on all sides until they were finally destroyed by Chief Seattle's Suquamish in the 1860s. Western civilization gained a firm foothold in Quileute country in 1882 with the establishment of a government school. Treaties were signed with the Quileutes in 1885 and 1886. Today a rebirth of cultural awareness among all the tribes on the peninsula has led to the teaching of native languages and tribal traditions to young tribal members, keeping ancient legacies alive as the last native speakers approach the end of their days.

The Quinault and the Queets

South of the Quileute are the Quinault and Queets people. Like the S'Klallam along the strait, and the Twana of Hood Canal, the Quinault are Salish-speaking people. Salish speakers probably spread south from the Fraser River delta and now inhabit much of Puget Sound and British Columbia's Georgia Strait, as well as the south Olympic Coast. The Quinault lived in five large winter villages near the mouth of the Quinault River; the closely related but less numerous Queets people lived in one or two villages on the nearby Queets River. Both occupied nearly 50 smaller settlements scattered along the two river systems — some a good ways inland of Quinault Lake.

Their territory was particularly productive; the Queets and Quinault rivers supported all five species of Pacific salmon, including large runs of sockeye in Quinault Lake, as well as anadromous steelhead trout. Sockeye were particularly important to the Quinault. They are rich in oil and flavor, and since they do not spawn in large quantities elsewhere on the peninsula, they were a cherished trade commodity. The numbers of sockeye and other salmon running in the Quinault River system may account for the large number of villages (and the large population) recorded there around the time of contact.

Like Coast Salish people elsewhere in the Northwest, the Quinault maintained numerous fish weirs and traps on their rivers. Each village had at least one weir, a "V" of sticks planted in the riverbed and woven with a mesh of other sticks or a net to trap or concentrate salmon or steelhead on their upstream migration. The fish could then be taken with dip nets or spears. A large run of sockeye that spawn in Quinault Lake in late winter and early spring provided a surplus that was dried for later use as well as traded. Early runs of smelt on the coast were taken for trade; the oil they yielded was highly valued on the peninsula. Halibut, cod, herring, and other ocean fish filled out the Quinault's seasonal fishery.

Like most Northwest Coast people, the Quinault practiced a First Salmon ceremony. Its intent was to honor and give thanks to the salmon, supernatural humans who lived in longhouses beneath the sea. If the salmon people were shown the proper respect and gratitude when they visited a village in salmon form, they would return in great numbers, bequeathing the gift of their flesh to the villagers, thus insuring the people's survival. For a First Salmon ceremony, the first salmon caught by the village headman in late winter or early spring was laid carefully on the river bank, its head pointing upstream. The fish was cut in a prescribed manner, using a mussel shell knife. The salmon's heart was carefully removed and ceremonially burned in a fire; its head, bones, and entrails were kept intact. After it was cooked, pieces of the first salmon were shared with each member of the village, and its bones were carefully returned to the river.

Salmon were central to all Quinault villages, though inland settlements probably also relied on elk, deer, and bear. Upriver Quinault are reported to have dressed in hides, furs, and moccasins and hunted with bows. Log "deadfall" traps were set for bears, and snares were used for smaller animals. Seasonal hunting forays took many coastal people into the high country. Traditions tell of inland skirmishes between Quinault and S'Klallam people who entered the mountains from the north. The notion that prehistoric Native American people avoided the mountains was laid to rest by the Olcott findings of the 1980s and 1990s. Accounts from tribal elders from around the peninsula indicate that this was just as untrue in recent times. Tribal traditions were once more proven accurate in the summer of 1994 when a low winter snowpack melted back on a high northern ridge to reveal a woven burden basket flattened against the newly exposed ground. Its style matches that of later Northwest Coast culture and it could easily date from the time of contact, indicating that high-country gathering remained a vital part of the seasonal round on the peninsula well into historic times.

The S'Klallam

The historical territory of the S'Klallam people comprised those northern ridges where the basket was found, the forested streams and rivers that drain them, and the 80-mile stretch of bays, points, and shoreline that form the peninsula's northern coast. At the time of European contact, about 1,500 S'Klallam occupied 13 winter villages along the strait from the Hoko River east to Discovery Bay. Small clusters of gable-roofed longhouses lined the shores of protected coves and river mouths. Large settlements were located at the mouths of the Clallam, Elwha, and Dungeness rivers, and at least one permanent village was located up the Elwha, a site now flooded behind Elwha Dam. With salmon spawning in the northern streams and all the Puget Sound runs passing through the strait, the S'Klallam enjoyed a readily available and nearly year-round source of protein. From late winter through spring they trolled and netted Chinook and coho in the strait. Halibut were plentiful in late spring, and sockeye and pink salmon arrived in summer. The S'Klallam shared the use of salmon camps of woven mat huts with their eastern neighbors, the Twana, along Hood Canal as well as with other Salish-speaking groups across the strait. Later in the season, as fall chum and winter steelhead runs headed up their rivers, the S'Klallam rounded out the year fishing their own streams with weirs and spears. Typically, village headmen controlled the first weir on each stream. Great quantities of salmon — as many as several thousand a day at the height of the runs — were dried on racks or smoked and stored.

S'Klallam people netted seabirds and ducks with large nets stretched over sand spits, or hunted waterfowl at night from canoes using long-handled dip nets. Each village had a deer and elk hunter and a sea mammal hunter. Furs, bones, and antlers were always in demand, and hunters and their families sometimes spent whole seasons in the mountains hunting and then dressing hides. Seals were hunted for meat and oil; whales were taken only occasionally, when one was sighted in the strait. Camas, bracken fern root, and other plant foods were gathered at Sequim, Port Townsend, and other prairies.

Like their coastal neighbors to the west, the S'Klallam were fierce warriors. They warred with the Makah as well as with tribal groups inhabiting Puget Sound. They in turn were raided by northern tribes, and many of their villages were well stockaded. Nonetheless, they frequently intermarried with Makah and Twana people and enjoyed extended social relations with other tribal groups, trading, gambling, and potlatching well up the coast. By the mid-19th century the S'Klallam had expanded their territory to the east, occupying the land formerly held by the Chemakum.

The Twana

Hood Canal, which bounds the eastern shore of the peninsula, was misnamed by cartographers. The "canal" is actually a glacial fjord carved into the base of the mountains. Fed by the rivers pouring off the steep eastern front of the mountains, and regulated by a glacial sill at its mouth, Hood Canal provides a rich marine environment hosting salmon, sea mammals, and exceptionally productive shellfish beds. Before the coming of Europeans, its shorelines, bays, and upland valleys were home for the Twana people.

Some 13 Twana villages have been identified along the shores and river valleys abutting Hood Canal. Villages were located at the mouths of the Dosewallips, Duckabush, and Skokomish rivers, as well as on the shores of Dabob and Quilcene bays and many smaller creek mouths. Most populous among the Twana were the Skokomish people, who lived in five or six villages along the lower Skokomish River and maintained a permanent village on the inland tributary, Vance Creek. The Twana villages along the canal shared a salmon-oriented culture similar to that of their S'Klallam neighbors, with perhaps a greater reliance on shellfish. Littleneck, butter, and horse clams, cockles, geoducks, mussels, and native oysters were gathered in quantities (Quilcene Bay remains one of the most important oyster-producing areas in the country).

Like the S'Klallam, most Twana villages had land mammal and sea mammal hunters, as well as those specializing in waterfowl. But the Skokomish villagers evolved some unique cultural patterns. The inland villages of the Skokomish were apparently more cohesive and closely allied than those of the coastal Twana. And while they relied heavily on salmon, land mammals were a key part of their local economy. In fall the villages would band together for organized elk hunts in the upper valley of the South Fork Skokomish. The headman of the most highly ranked village led the hunt, assisted by specialists whose spirit power gave them influence over the movements or behavior of elk.

As with other activities that involved the taking of animals, elk hunting was carried out with elaborate ritual and preparation. The Skokomish practiced a First Elk ceremony as well as a First Salmon ceremony. Families of hunters dressed the hides and processed and cured the meat over low fires. The meat was temporarily stored in raised cedar caches and later moved to winter villages. Vital food-procuring and tool-making skills among the Skokomish — as among other tribal groups on the peninsula — were often believed to have come from a spirit power. From an early age, young people were encouraged to make a spirit quest into the forest, to fast and bathe in hope of being visited by a spirit power who would assist and guide them later in life.

Coast people believed in a mythic time when creatures shared both human and animal characteristics. To prepare the world for humans, the Changer transformed many of these creatures into the animals we know today. Some frightful beings were changed into rocks or landforms; others were banished to the mountains or the depths of the ocean. People were given their human shape then, and instructed in the fundamentals of culture. During a spirit quest, a youth or adolescent might be approached by an animal or spirit being and given a song. The seeker might then fall into a trance, enter myth time, and be taken to his or her spirit animal's human village, where the youth would be given special powers. Among the Twana, it wasn't until years later in an individual's life, usually during an illness, that a shaman would recognize the return of a spirit power and assist in a ritual where the spirit song would be sung, danced, and acknowledged among the village, perhaps to be melded into the cycle of winter ceremonies.

These strong ritualized ties with the world of spirit permeated every aspect of native life. Every skill, every success in hunting, fishing, or gathering was accomplished only with help of guardian spirits. By aligning themselves with the spirit powers that governed the abundance and availability of plants, fish, birds, and other animals, and by following the mores prescribed for taking them for their use, indigenous peoples lived as equals within their larger natural communities. The means of sustaining a human population through times of scarcity and abundance, disturbance and environmental change were woven into the fabric of native cultures over millennia and reinforced through a strict set of beliefs and ritual practices. The fit between these primary cultures and the changing dynamics of the Olympic ecosystem was seamless.

Treaties and Treaty Rights

By the mid-19th century, when Isaac Stevens was appointed governor of the newly created Washington Territory, the fabric of Native American culture had been deeply torn, and remaining populations were ravaged by disease and alcohol. The Donation Land Act of 1850 had allowed settlers to stake claim to Indian lands well in advance of negotiated treaties. The years 1851 and 1852 saw the first registered homestead claims in the Port Townsend and Sequim areas. Gold was discovered in eastern Washington Territory a few years later, and the floodgates opened. Liquidating native land claims to make way for settlers was high on Governor Stevens's agenda, and he accomplished this through a series of hastily drawn treaties. Stevens signed treaties in 1855 and 1856, negotiating with designated "tribes" and appointed "chiefs" in Chinook jargon, a trading medium of limited vocabulary wholly inadequate for complex treaty issues. Nonetheless, Stevens persuaded Native Americans to cede ownership of their lands (a concept foreign to native cultures) in exchange for reservations, certain specified goods and services, and the guarantee that native people would retain the right to hunt and gather on open and unclaimed land and to fish "in all their usual and accustomed grounds and stations" in common with whites. This last part of the agreement proved critical over the next century and a half as tribes challenged a relentless array of state-imposed assaults on their fishing rights.

In 1905 and again in 1908, Northwest tribes won important Supreme Court contests affirming their right to fish in accustomed areas. Nonetheless, by the 1960s, Washington state agents were arresting native fishermen on the grounds that traditional net fishing was "incompatible" with modern fisheries management. Although the tribal portion of the statewide catch amounted to only about 6 percent at the time, natives were blamed for the overfishing that was depleting the runs.

Following a 1964 attempt by Washington state to nullify tribal fishing rights on the Nisqually River, the federal government again stepped in. In a landmark 1974 decision, U.S. District Court Judge George Boldt upheld treaty fishing rights and changed over a century of fisheries management in Washington. Judge Boldt interpreted the 1855 treaties to mean that Indians were entitled to half of the harvestable salmon and steelhead in Washington state. A later decision granted tribes the right to protect the habitats of salmon and steelhead. That decision led to the 1987 Timber, Fish, and Wildlife Agreement, which gives tribes, state agencies, timber companies, and environmental groups an opportunity to work together to find solutions to threats to salmon habitat. The TFW agreement places Native Americans' traditions and ethics once more in the forefront of habitat protection on the peninsula, a role tribal peoples have assumed for hundreds if not thousands of years.

A Cultural Reawakening

Native Americans have been remarkably resilient in maintaining cultural traditions in the midst of profound economic, social, and religious change. Largely through oral traditions, elders have managed to keep tribal identities and values intact despite intensive government effort to "assimilate" native peoples into American culture. This resilience, combined with federal acknowledgement of long-suppressed treaty rights, and the rediscovery of a rich and prideful heritage, has led to a cultural reawakening among Native Americans on the peninsula and throughout the Northwest coast. Programs to preserve native languages and teach them to tribal members, begun as early as the 1960s and 1970s among the Quileute and Makah, are now well underway among other peninsula tribes. And potlatches and tribal gatherings remain vibrant with traditional songs and dances.

Perhaps the most visible expression of this rebirth of Native American culture on the peninsula has been the resurgence of the redcedar canoe. To commemorate the centennial of Washington's statehood in 1989, 16 Washington tribes took part in a celebratory "Paddle to Seattle." Several tribes carved traditional canoes from redcedar logs for the occasion. At a welcome celebration hosted by the Suquamish Tribe, the paddlers were invited to participate in a much more ambitious international effort, a Paddle to Bella Bella on British Columbia's northern coast.

Four years of planning, carving, and training led up to the event, and in the summer of 1993, six Washington tribes and an international group pointed their cedar canoes north for the 650-mile journey and began singing their paddling songs. The Quinault, Quileute, Makah, and all three S'Klallam tribes mounted traditional canoes, crews, and support boats. The enthusiasm and dedication of the paddlers and the spirit of intertribal cooperation proved infectious. Whole families of supporters, from elders to small children, went along in cars, trucks, and vans. At each stop along the way, paddlers and guests were ceremonially welcomed and hosted by local tribes. Singing, dancing, and feasting filled traditional longhouses along the northern coast as nations celebrated their diversity as well as their common cultural bonds.

Many Native Americans on the peninsula see the Paddle to Bella Bella as a turning point in cultural awareness for tribal members of all ages. Elwha S'Klallam carver and head paddler Al Charles Jr. was 21 at the time of the paddle. "To be able to see our culture and beliefs put into practice on a daily basis really brought the culture alive for young people," he said later. "The farther we paddled toward Bella Bella, the more people came out of their shells, joining in the singing and drumming. By the north end of Vancouver Island, we were singing our tribal songs for crowds of 300 to 400 people as we pulled into the beaches."

For Al Charles Jr. and for tribal members across the peninsula and Puget Sound, the change was lasting. "I hear young people on the reservation just break into a song now, something they'd be afraid to do before the paddle," he told me. "And the experience has put a twinkle back into the eyes of our elders."

10
Protection for Olympic's Wildlands

In 1890, the same year the U.S. census pronounced the American frontier closed, a "rush" of exploration hit the Olympic Mountains. Two of the Olympics' most noteworthy explorers met in July of that year at a forest camp in the Skokomish Valley. Judge James Wickersham was an inveterate explorer and mountaineer who had traveled widely in the Northwest. A year earlier he and a companion explored the upper reaches of the Skokomish Valley; he returned this year with a larger party to pioneer a route north across the mountains to Port Angeles.

Sharing a campfire with Wickersham was Lieutenant Joseph P. O'Neil. Also a seasoned wilderness traveler, O'Neil had led an exploratory expedition into the northern Olympic high country five years earlier, eventually reaching the Dosewallips headwaters and the northern flanks of Mount Anderson. "The travel was difficult," he confessed in his report, "but the adventures, the beauty of the scenery, and the magnificent hunting and fishing amply repaid all hardships." Now he returned with a large party of enlisted men and members of the Oregon Alpine Club, a pack train, and mountains of supplies. They were building trail as they advanced up the Skokomish River, and progress was slow.

Janis Burger

Mount Carrie and the Cat Basin. Though the interior mountains had been traversed by Native Americans for millennia, they were terra incognita *for Euroamericans until the latter part of the 19th century.*

Though members of O'Neil's party would explore sections of the Duckabush and Dosewallips rivers to Hood Canal and descend the East Fork Quinault River to the Pacific that summer, O'Neil's men were not the first to cross the range. A small party led by Melbourne Watkinson is reported to have crossed by a route similar to O'Neil's in 1878, but the glory of "first Olympic crossing" went to a well-publicized expedition sponsored by the *Seattle Press.* Its ragtag members emerged from the rain forests of the Quinault, battered, footsore, and hungry, only weeks before Wickersham and O'Neil's meeting. The Press party had started across the Olympics from the north the previous December, largely in an effort to scoop Wickersham and O'Neil (few were then aware of the Watkinson expedition) and ran headlong into one of the worst Olympic winters on record. The travels and travails of the Press party's ill-timed adventure are delightfully told in Robert L. Wood's *Across the Olympic Mountains.*

No record remains of the conversation that Wickersham and O'Neil shared around the fire that July afternoon. They undoubtedly discussed their previous routes and the nature of the country they had crossed. It's also quite likely that there, as afternoon breezes wafted their smoke through the valley forest, the seed for the creation of Olympic National Park was planted.

By the late 1800s, the wholesale destruction of the virgin pine forests of the Great Lakes states by large timber companies sparked a nationwide movement to protect American forest lands. In 1864, a federal grant allowed the State of California to designate Yosemite as a park. In 1872, President Grant signed the act creating Yellowstone National Park, and by 1885, nearly three quarters of a million acres were set aside as a forest preserve in the Adirondacks in New York. The popular writings of John Muir extolled the majesty of western mountains and forests and demanded government protection for them. In 1891, Congress responded to growing popular sentiment and gave the president power to create forest preserves by proclamation, withdrawing land from the public domain and closing it to homesteading and private speculation. Judge Wickersham knew that Major John Wesley Powell, head of the U.S. Geological Survey, was preparing a list of forest reserve recommendations for the president's signature. Wickersham sent Powell his unpublished report and urged that Olympic be recommended as a national park:

> In all the great commonwealths of Idaho, Montana, Nevada, Oregon, Washington and Alaska, there is not a national park. The forests are being felled, and destroyed, the game slaughtered, and the very mountains washed away, and the beauties of nature destroyed or fenced for private gain. . . . The heaviest growth in North America lies within the limits of [the Olympic] region, untouched by fire or axe, and far enough from tidewater that its reservation by the government could not possibly cripple private enterprise in the new state, and by all means it should be reserved for future use. The reservation of this area as a national park will thus serve the twofold purpose of a great pleasure ground for the Nation, and a means of securing and protecting the finest forests in America.

O'Neil was equally struck by the beauty of the peninsula's mountains, forests, and streams, and the abundance of wildlife he found there. He concluded his report, written in 1890 and published six years later, by noting that while the area was unfit for settlement, it would "serve admirably as a national park."

But the commercial wealth of the Olympics' forests did not go unrecognized by powerful timber interests. It would take nearly half a century — and one of the most bitterly fought conservation battles in American history — before Wickersham and O'Neil's vision of a national park in the Olympics would become a reality.

Early Protection for the Olympics

Following the passage of the 1891 Forest Reserve Act, President Cleveland appointed a committee to make recommendations regarding proper management of the forest reserves. Two key figures in American conservation were involved in the committee's deliberations: John Muir, an eloquent and impassioned advocate of forest and wildland preservation and founder of the Sierra Club, and a young millionaire named Gifford Pinchot, a leading proponent for the utilization of public forests. Muir envisioned a system of reserves protected from logging, grazing, mining, and dams, in the manner of Yosemite and Yellowstone. Pinchot, with his background in European forest management, saw the reserves quite differently. Thus began a national debate over the use of public forest lands that still rages in the Pacific Northwest a century later, a debate in which the Olympic Peninsula has served as center stage.

Though Pinchot's influence held sway over the committee's recommendations, its final report recommended the establishment of Mount Rainier and Grand Canyon national parks. The committee also recommended a large forest reserve for the Olympics. Established by President Cleveland on February 22, 1897, the 2,188,800-acre Olympic Forest Reserve encompassed all of the mountains, foothills, and lowland valleys of the peninsula, along with the rich, productive forests of the western slope out to the Pacific coast. It was by far the largest as well as the most valuable forest reserve in the nation, and timber interests in Washington state lost no time mounting a campaign to wrest it from public hands.

It took three years to complete the considerable task of surveying the new reserve and assessing its forest resources. That was all the time commercial interests needed. Thomas Aldwell, a Port Angeles real estate speculator who would later become the developer of the Elwha Dam, petitioned Congress to eliminate the most valuable timber lands from the reserve. He predicted economic ruin for the peninsula, going so far as to invoke the indebtedness of "many widows and orphans in the East who had been persuaded to buy securities" in Clallam County. Congressman William Jones fanned the flames with rhetoric, crowing that the reserve's establishment was "the most despotic act that has ever marred our history." Local officials maintained that the peninsula's rain-drenched coastal plain, remote from population centers and crowded with some of the largest trees in the world, was too valuable as agricultural land to be locked up in a forest reserve. They pointed to the public law stating that parts of any reserves "better adapted for mining or agricultural purposes than for forest usage may be restored to the public domain." Demands by local politicians, chambers of commerce, and industrialists were well received in Washington D.C., particularly after a timber company executive was installed in the United States Senate.

In 1900, a proclamation by President William McKinley reduced the reserve by more than a quarter-million acres; the following year he cut it by nearly a half million more. By the time the government survey of the Olympic Forest Reserve was complete — a report specifying that soils, rainfall, and the size of the forest made the area completely unsuitable for agriculture — three-fourths of the richest forest lands in the reserve had been opened to claims by settlers.

Chris Morgenroth homesteaded in the Bogachiel Valley a decade earlier and developed many of the first trails on the western peninsula. After McKinley's deletions, he watched in dismay as "homesteaders" made fraudulent claims to the best tracts of forest, then turned them over to timber company agents for a tidy profit. Contrary to the claims of Aldwell and company, this mass transfer of public forest land into the hands of a few private timber companies dashed the hopes of Morgenroth and other early settlers for rural development of the area. Among the new "settlers" in the Olympic rain forest valleys were the Milwaukee Land Company, Weyerhaeuser, Simpson Logging Company, Merrill and Ring, and others. More than a half-million acres of some of the nation's finest forest lands passed, in the words of Gifford Pinchot, "promptly and fraudulently into the hands of lumbermen."

While this first great land grab was underway, another sort of trade was devastating the Peninsula's elk herds. Elk carcasses were found scattered across the western valleys stripped of their hides and incisor teeth. The teeth were popular among members of the Benevolent and Protective Order of Elks as watch fobs, and commercial hunters killed elk indiscriminately. By the turn of the century, the elk population for the peninsula had been reduced to less than 2,000 animals. The *Seattle Mail and Herald* published a lengthy exposé on the situation, and conservationists around the state clamored for protection. In 1905, the Washington state legislature passed a stopgap 10-year moratorium on elk hunting on the peninsula, a measure legislators hoped was a "first step in what will ultimately be Olympic National Park."

In 1906 Congress passed the Antiquities Act, giving the president the

The Queets Fir in Olympic National Park. More than three-quarters of a million acres of some of the most productive low-elevation forests on the Olympic Peninsula were fraudulently removed from the Olympic Forest Reserve in 1900 and 1901. Nearly all of it has since been logged.

Olympic National Park

power to preserve objects of historic or scientific interest by designating national monuments by proclamation. Conservationists saw this as an opportunity to bypass a Congress heavily influenced by timber interests and gain federal protection for the Olympics. The following year, the Mountaineers, a Seattle hiking and climbing club, organized a large outing into the heart of the reserve. Its members, moved by the beauty of the forests and glacier-clad peaks, enthusiastically took up the cause of a national park. During the last days of his presidential administration Theodore Roosevelt, an early champion of conservation causes, was asked by Congressman W. E. Humphrey of Tacoma to create a national monument in the Olympics. Roosevelt (for whom the Olympics' elk were later named) was agreeable, and on March 3, 1909, he designated the 600,000-acre Mount Olympus National Monument.

The monument protected only the upper portions of the rain forest valleys remaining in the forest reserve and virtually none of the winter habitat critical to elk, but it was a start. Together with the coastal wildlife refuges Roosevelt had established two years earlier, the groundwork for the protection of the Olympic ecosystem was in place.

The Battle for a National Park

In his excellent book, *Olympic Battleground*, Carsten Lien traces the long campaign for a national park in the Olympics. In one sense, it was a competition between two youthful land management agencies — the forest service and the park service — for the same lands. Gifford Pinchot had been appointed to head the Department of Agriculture's Division of Forestry. After the timber giveaways and other abuses of the forest reserves in the early 1900s, he succeeded in having the reserves transferred from the Department of the Interior to the Department of Agriculture in 1905. And so the Olympic Forest Reserve became Olympic National Forest.

Pinchot imbued his agency with an ethic of commercial utilization and responsiveness to the needs of local resource industries. John Muir and the growing number of preservationists of the day had no illusions that the reserves would be managed for anything but timber under Pinchot's new agency; they pressed even harder for the creation of new national parks. Since many of these parks were carved out of existing national forests, the fundamental dichotomy within the American conservation movement was cast in stone. The National Park Service, created at the urging of preservationists in 1916, was directed to manage its lands "to conserve the scenery and the natural and historic objects and the wildlife therein and to provide for the enjoyment of future generations." In order to build a constituency for national parks by increasing their popularity, the park service's first director, Stephen Mather, emphasized the "enjoyment" clause in his mandate. The roads, resorts, golf courses, and entertainments he brought to Mount Rainier and other newly designated parks increased their popularity but lost critical preservationist support in the Northwest at a time when Olympic needed it most.

By 1912, a coalition of Washington timber and business interests recommended removing all stands of commercially valuable timber from Mt. Olympus National Monument, then designating the remaining high country a national park — albeit one that would allow logging, mining, and dams. This was a strategy that timber companies as well as the U.S. Forest Service would pursue throughout the ensuing quarter-century-long battle for a national park. With the election of President Woodrow Wilson, the timber companies got their wish. In 1915, Wilson slashed 170,000 acres of forest from the monument, eliminating all that was left of the rain forest valleys and virtually all of the heavily timbered southern Olympics. Park bills in 1911 and 1916 were also defeated in Congress. But local and regional sentiment for a national park continued to build, and the Mountaineers and other conservation and outdoor groups continued to press for a park.

During World War I, the Army mounted a major effort to harvest the dense stands of Sitka spruce on the west side of the peninsula; the clear, lightweight wood was needed for the construction of warplanes. The Army's Spruce Production Division took on the massive undertaking of building a 36-mile railroad to deliver spruce logs to Port Angeles. The railroad was the fastest (and most expensive) ever constructed in the United States, but the Armistice was signed a few weeks before the railway was complete. It did, however, provide a means for timber companies to log the western rain forests. Lumbering now began in earnest.

By the late 1920s, much of the forests deleted from the original

Olympic Forest Reserve were gone, and mill owners turned their gaze to the trees in the national forest. The forest service was happy to accommodate, and cutting levels on Olympic shot from around 12 million board feet in 1915 to 175 million by 1928. In its 1926 timber harvest plan, the forest service went so far as to include forests inside the national monument in its cutting circles. Around this time, citizen efforts for a national park took on a nationwide scope.

Continuing concern over the fate of Roosevelt elk in the Olympics prompted the Izaak Walton League to become involved in the park

Lake Crescent circa 1912. As roads were opened into the Peninsula's lowland forests and visitors encountered Olympic's magnificent old-growth trees, sentiment for preserving a part of this heritage as a national park began to grow.

Bert Kellog Collection, North Olympic Library System

issue. And in New York, an assistant curator of the American Museum of Natural History named Willard Van Name emerged as a key player. Respected among the scientific and conservation communities, Van Name was a sharp and widely published critic of both forest service and park service management of public lands. In 1930, Van Name joined forces with Irving Brant, a Saint Louis newspaper editor, and Rosalie Edge, an outspoken New York conservationist active with the Audubon Society. They formed the Emergency Conservation Committee (ECC), and protection of the Olympics' magnificent old-growth forests was high on their list of priorities. In 1932, Van Name traveled to the Olympic Peninsula to investigate the situation. He reported that none of the lands within Olympic National Forest offered secure protection for either forests or wildlife, nor were roads or summer homes prohibited in the national monument. He also learned that the recently completed Olympic Highway would allow the forest service to intensify its cutting of the national forest. Van Name was particularly struck by the grandeur — and vulnerability — of the western rain forests, and he returned to the East determined to fight for their protection. Van Name and the ECC proposed a large park, which would include portions of the forested valleys of the Sol Duc, Bogachiel, Hoh, Queets, and Quinault, as well as Lake Crescent and the north shore of Quinault Lake.

Through its publication and wide distribution of pamphlets and papers, the ECC was successful in galvanizing members of national conservation organizations, scientific societies, universities, and natural history museums into an informed and effective broad-based coalition. Through letters to newspapers, agencies, congressmen, and the White House, citizens expressed overwhelming support for an Olympic National Park. With the election of Franklin D. Roosevelt in 1933, the campaign for a park hit high gear. Van Name stated his case forcefully:

> The Peninsula affords the last opportunity for preserving any adequate large remnants of the wonderful primeval forests of Douglas fir, hemlock, cedar, and spruce which were not so many years ago one of the grandest and the most unique features of our two northwestern-most States, but which everywhere have been or are being logged off to the very last stick.

With the movement for a park reaching a boiling point, the park service now conducted its own study into the feasibility of making Olympic a national park. Cowed by the forest service, the park service rejected the ECC's proposal, claiming it included "much available timber on lands that are not of national park caliber" in the proposed park. It recommended instead a much smaller park devoid of commercially valuable timber.

When the park service distributed a copy of its draft report to a timber industry lobbying group, Van Name went directly to Roosevelt's Secretary of the Interior, Harold Ickes. At Ickes's direction, the park service revised its recommendations to agree with the conservationists' proposal. The resulting bill was opposed, of course, by the forest service and the timber industry, and failed to pass. Pressure ran high, and the two sides were deadlocked. It was then that President Roosevelt entered the discussion, determined to resolve the issue. He traveled to the Olympic Peninsula to look into the matter for himself.

When Roosevelt arrived in Port Angeles on a drizzly day in September 1937, he was met by more than 3,000 schoolchildren from around the peninsula, who turned out in front of the Clallam County Courthouse beneath a banner that read, "Please Mr. President, we children need your help. Give us our Olympic National Park." Popular support for a park was at its peak on the peninsula at that time, and Roosevelt told the children and an assembled crowd of about 10,000 adults that they could count on his help in getting a park for the Olympics.

That evening, in his cabin at Lake Crescent Lodge, Roosevelt made it clear to representatives of the park service, the forest service, and the congressional delegation that he wanted a large park, with "more large timber" than either agency had proposed. Roosevelt concluded his visit with an auto tour of the western peninsula, stopping at Forks and Lake Quinault Lodge. In a comically futile gesture, the forest service — clearly the loser in the president's visit — had moved the sign marking the southern boundary of the forest two miles north in an attempt to disown a large fire-blackened clearcut.

FDR's trip crystallized national attention on the park, defeated forest service opposition, and at least temporarily silenced the timber industry. On June 29, 1938, Roosevelt signed the bill creating Olympic National Park. The park encompassed 638,280 acres, with the provision that the president could expand it by proclamation up to 898,292 acres. In order to insure protection for the peninsula's coastal strip, the Bogachiel rain forest, and a corridor along the Queets River within the allowed acreage (all were high priorities for FDR), conservationists reluctantly agreed to eliminate the the middle Dosewallips Valley and much of the upper Dungeness watershed in the northeast Olympics.

At long last, the battle to create Olympic National park was won. Some of the finest examples of old-growth forest and wildlife populations in the Pacific Northwest were preserved for future generations. It had been a long and contentious struggle. A powerful industry, one that had controlled the political life of the region for nearly a century, and a timber-oriented government agency had been overcome by widespread citizen activism and the vision of a handful of dedicated leaders. At a celebratory banquet in Seattle, Secretary Ickes reaffirmed that Olympic would remain a wilderness park. But any savoring of the victory on the part of conservationists would prove short-lived. Though the park was established, the battle to preserve its irreplaceable old-growth forests was far from finished.

Attacks on the West-Side Forest

With the onset of World War II, the timber industry redoubled its efforts to remove valuable west-side forests from the park. In 1944, to the horror of the environmental community, the park service recommended deleting from the park much of the lower Bogachiel as well as lowland rain forests in the Quinault Valley. When these recommendations were introduced as a bill in 1947, the same environmental coalition that brought about the creation of the park rose to its defense. Letters poured in to congressional offices and to the park service condemning the bill. The public won this battle, but conservationists realized that protection of Olympic's forests would require continued vigilance. Irving Clark of the Mountaineers organized the Olympic Park Associates as a "watchdog" group dedicated to preserving the wilderness integrity of the park. OPA and other conservation organizations had their work cut out for them as assaults on the newly created park continued.

In the meantime, lumbering interests on the peninsula found a way to quietly move old-growth trees out of the park and into local mills by means of "salvage" contracts. The 1916 Organic Act creating the National Park Service allowed for the removal of insect-damaged trees to preserve forest health. At Olympic, this directive was broadened to include the salvage logging of downed and damaged trees as well as any *potential* insect trees. Beginning in 1941 and continuing until it was stopped by a national outcry from environmental organizations in 1958, more than 100 million board feet of timber, both standing and down, was removed from the park and trucked to area mills.

Other threats to the park, including several proposed roads that would have cut through fragile alpine meadows and opened the wilderness coast to traffic, were soundly defeated. A final effort to delete the Bogachiel River valley from the park was mounted as part of a "boundary adjustment bill." Ingeniously, it was packaged with legislation that would create North Cascades National Park in Washington. Again it was defeated by citizen conservationists. The history of preservation in the Olympics has been one of concerned citizens making their voices heard.

It wasn't long after these episodes that I unwittingly strolled into this citizens' campaign. On a hike along Shi Shi Beach in the early 1970s, I followed the sound of chainsaws up a creek and stumbled into a clearcut that seemed to extend for hundreds of acres. This northern reach of coastline was not included in the park at that time. Climbing past stumps and logging slash to the top of a rise, I looked down on timber fallers working a stand of spruce in a shallow draw. Trees dropped like matchsticks. It stunned me to discover logging like this no more than a 20-minute walk from one of the most startlingly beautiful beaches I had ever seen.

Not long after that I found myself working with OPA and other environmental groups to secure protection for the splendid northern coast. Protection came in 1976 when Congress added Shi Shi, Point of the Arches, and the east shore of Lake Ozette to Olympic National Park. By that time, virtually all of the Ozette basin had been logged, and forest service and Washington Department of Natural Resources clearcuts were chewing back the old-growth forests flush with the park boundary. The loss of cover for elk, spawning areas for salmon, and lowland old-growth habitat were affecting park resources. It became clear to me and other conservationists that if Olympic National Park was to remain ecologically viable, some means of protecting critical ecosystem functions on lands outside the park would be needed.

Protection for the Olympic Ecosystem

Nowhere has the inadequacy of Olympic National Park's boundaries proved more apparent than in the high rugged country of the east Olympics. Here the park boundary cuts randomly across alpine areas, migration routes, and the upper portions of six river drainages. Calving and wintering areas for east-side elk herds remain unprotected, and critical habitat for stocks of native anadromous fish have been lost to logging and development. As clearcutting accelerated on the national forest in the decades following World War II, roads were punched into upper watersheds, and steep mountainsides wholly unsuitable for timber management were clearcut, often with disastrous results.

Following passage of the 1964 Wilderness Act, national forests were directed to inventory remaining roadless lands and make recommendations for wilderness designation. Much of the high scenic country adjoining the east side of the park was recommended by the agency, including most of the botanically rich alpine areas of the Dungeness rainshadow. But few of the lower forested valleys were recommended, and very little of the heavily timbered west side of the forest. Out of approximately 53,000 acres of roadless lands inventoried on the two west-side districts of Olympic National Forest, only part of one area, a 12,000-acre ridge above Quinault Lake, was recommended for wilderness.

In the early 1970s, local conservationists began a grassroots campaign to rally support for protection of ecologically important forest service lands surrounding the park. A decade-long effort culminated with the passage of the 1984 Washington Wilderness Act. The Buckhorn, Brothers, Mount Skokomish, Wonder Mountain, and Colonel Bob wilderness areas were designated by Congress, protecting some 90,000 acres of outstanding wildlands. Important migratory and calving areas for east-side elk and critical salmon habitat along the lower Gray Wolf and the middle Dosewallips and Duckabush river valleys were included.

Congressional designation of forest service wilderness areas was an important step in establishing an interagency network of protected habi-

Janis Burger

tats on the peninsula. But as overcutting on federal, state, and private lands accelerated during the 1980s throughout the region — leading eventually to legal action mandating protection for the spotted owl — scientists and conservationists sought new ways to address habitat needs on an ecosystem scale. When a bill was introduced in 1988 to include 95 percent of Olympic National Park's lands within the national wilderness system (fulfilling the promise of Secretary Ickes a half-century earlier), it contained a section calling for an ecological study of all three of Washington's national parks. The five-year study would have assessed the parks' abilities to preserve ecosystems and "maintain healthy populations of all species native to the parks." The wilderness bill passed; it added the intertidal area to Olympic's coastal strip, including the wildlife refuges of the offshore rocks and islands. But the ecosystem study was seen as a threat by timber interests and was dropped from the bill.

One of the last unprotected treasures of the Olympics, the lower Gray Wolf River canyon was among some 90,000 acres of national forest lands surrounding the park that Congress added to the national wilderness system in 1984.

A central issue facing protected natural areas worldwide is their long-term viability. The dynamics of natural change in a system are compounded by the effects of human activities. Though humans have lived in the Olympics for the past 12,000 years, the ecosystem has only recently been subjected to the impacts of multinational timber operations and unchecked urban growth. Conservation biologists question whether our natural areas and preserves can continue to provide viable habitat under these pressures. Most feel this is possible only with sensitive interagency management of the larger ecosystems that surround protected areas.

Ecosystem management means managing large areas to insure that all plants and animals are maintained at viable levels in their natural habitats and that basic ecosystem processes are perpetuated over time. The ecological history of the peninsula since the retreat of the Vashon ice shows that this isolated, "islandlike" landform was not only adequate to provide viable habitat for a range of plant and animal species, it supported an exceptionally high level of biological diversity for an area its size. But the last 100 years of intensive human use of the peninsula has effectively shrunk that habitat island for many species to the confines of the park and protected areas within the national forest. The wolf is gone from the peninsula, and we may have lost the fisher as well. Native fish stocks continue to decline in some peninsula rivers, and concern over the fate of the spotted owl and other species dependent upon old growth continues. Though the peninsula's forest communities are incredibly resilient, the pressures now facing them are without ecological precedent.

A preliminary attempt to address ecosystem issues on the peninsula came with the Northwest Forest Plan. Released by federal agencies in 1994 in response to the biodiversity crisis in Pacific Northwest forests, the plan aims to provide critical habitat for old-growth-related species such as the northern spotted owl and to offer protection to streams and fish. The plan established a series of late successional forest reserves (old-growth and near-old-growth forest areas) and a network of riparian reserves and identified key watersheds on national forest lands for analysis and further planning. It reduced timber cutting on federal lands by about a third, but it failed to insure the survival of the region's old-growth-associated wildlife species. More than a third of these species have less than an 80 percent probability of survival under the plan.

Ultimately, the diverse natural systems of the Olympic Peninsula are parts of a single ecosystem sharing the same rich genetic reserves and biological processes. We are fortunate in the Olympics that many of the most sensitive and crucial parts of the ecosystem have been protected. Generations of Americans touched by the beauty and power of the Olympics have spoken up for its protection. But protection for the Olympic ecosystem is far from complete.

Continued deforestation of the peninsula's lowlands, degradation of salmon streams, loss of forestlands to residential development, and increasing hunting pressure on all sides of the park call for new approaches to conservation. In the face of such pressures, the island ecology that gave rise to much of the uniqueness of the Olympic ecosystem becomes its biggest liability. Genetic isolation places threatened species like the northern spotted owl in a particularly vulnerable position, and extirpated species like the gray wolf are not able to naturally recolonize the area as they have in the northern Rockies and North Cascades.

Keeping Olympic Wilderness Safe

If Americans are to complete the century-long process of protecting the Olympic wilderness for future generations, a number of tasks remain. Key lands need to be added to park and forest service wilderness areas, but such additions alone will not secure ecosystem protection. Removing the illegal dams on the Elwha River and restoring native salmon runs to this largest of the peninsula's watersheds should be a top priority. It is also essential that protection be extended to portions of existing free-flowing rivers that lie outside the park's boundaries. Several Olympic streams have been targeted for small-scale hydroelectric development, and logging and development continues within riparian areas. At present, not one of the peninsula's many wild rivers is included in the National Wild and Scenic Rivers system. The forest service has recommended portions of three rivers for designation; conservationists have proposed several more. It remains for Congress to take necessary action to complete this vital link in ecosystem conservation.

Another necessary step is restoring the Olympic's top predator, the wolf, to its former range. Wolf reintroduction is proving successful in the Yellowstone and northern Rocky Mountain ecosystems. Restoring wolves to the park would rekindle this predator's evolutionary relationship with Roosevelt elk, and return its haunting music to these coastal valleys and ridges.

With an eye toward the next century and beyond, scientists and resource managers will have to look beyond existing administrative boundaries. Managers will need to seek new and untried avenues of cooperation between administrative agencies, state, tribal, and local governments, as well as commercial developers, timber companies, and private landowners. As both the local population and the number of visitors to the peninsula continue to increase, it will be essential to establish cooperative measures and partnerships to secure seasonal wildlife easements through private lands, protected migration corridors, networks of integrated old-growth habitats, wetlands conservation strategies, and site-specific hunting and fishing regulations. Interagency efforts focusing on endangered species elsewhere in the West may provide working frameworks, but over the long term, specific legislation keyed to ecosystem management is needed.

For this approach to succeed, there is an immediate and pressing need for adequate research. An ecosystem study such as that proposed in 1988 would provide a comprehensive inventory of baseline species and habitat needs on the peninsula. Ecosystem process studies would determine how various wildlife species react to changes in forest composition, and ongoing research would feed information back to land managers, allowing the fine-tuning of resource management and restoration activities.

The world scientific community has twice recognized the importance of Olympic National Park on a planetary scale. In 1976 the United Nations Educational, Scientific and Cultural Organization (UNESCO) designated Olympic as part of its international system of Biosphere Reserves. The Man in the Biosphere program, of which the reserves are a part, identified representative examples of the world's major ecosystems in 72 countries. The reserves are intended to safeguard genetic diversity worldwide. They also serve as research centers and as benchmarks against which to measure human impacts on surrounding areas.

In 1982, Olympic National Park was selected by UNESCO as a World Heritage site. As such, Olympic takes its place among some of our planet's most treasured natural and cultural sites: the Grand Canyon, the pyramids of Egypt, the cathedral at Chartres. Such places are part of our common human heritage. UNESCO recognizes that these and other World Heritage sites are "older than the nation where they happen to be located. They belong in a way to the entire world."

The Estuary

Winter fog hugged the shoreline and spilled over the road as I wound my way south along Hood Canal in the early predawn light. As I dropped into the Dosewallips River floodplain and approached the narrow two-lane bridge, dark shapes rose through the fog, danced across the roadway, and slipped off into the mist. I killed the engine and rolled to a stop on the gravel shoulder. Stepping out of the car, I watched a herd of 30 or 40 elk bound off through stiff winter marsh grass, then spread over the estuary and begin to feed.

A few weeks earlier my wife, Mary, and I had wandered across this estuary and found side channels of the river full of spawning chum salmon. Their iridescent shapes rippled beneath the surface. Now, as the morning lightened and the fog began to lift, the blond rump patches of elk ranged out toward the slack blue-gray waters of the canal — mountain gods come to drink at the sea's edge. The summer before, I had watched what may have been these same elk browsing the high meadowlands below Hayden Pass in the long shadows of evening. The shrill calls of the bulls echoed off surrounding peaks. The rut now over, they had followed their old trails down the steep mountain valley, through open bottom lands and dense patches of second growth, fenced pastures and scattered farms, to slip past the small sleeping town and cross the highway into the deep growth of the estuary. It must have been an ancient route for them, carved when winter herds dappled the shores of these inland waters and salmon swelled the streams. We've left few valleys wild enough for these grand migrations. Still, the elk manage to return, their large hooves clacking across the blacktop.

An empty log truck rumbling by brought me back to myself. The lowlands had cleared, and clouds were rising against the steep forested ridges up valley. Mount Constance, one of my favorite winter views, stayed hidden. Snows deepened over the summer ranges of elk and deer, heather vole and marmot. Icy water churned over the deep redds of salmon and carried a plume of nutrients into the saltwater bay where grebes and mergansers wintered.

It occurred to me that the seasonal pulse at the heart of these mountains was not unlike the beat of our own hearts. It infuses the land with vitality, beauty, and grace. Maybe that is why human hearts have responded to this place so passionately. Maybe its future and our own are deeply joined.

General Information

Visitor Centers

The **Olympic Park Visitor Center** is located at 3002 Mt. Angeles Road in Port Angeles. It offers interpretive exhibits, a small theater/auditorium, a children's activity room, and a nature trail. The **Hoh Visitor Center** is located in the heart of the Hoh Rain Forest. It has exhibits and three short nature trails. Both offer a wide selection of publications as well as maps and information. Both are open all year, although the Hoh Visitors Center is staffed only from late spring to early fall.

The joint U.S. Forest Service/Park Service **information center in Hoodsport** offers information on the east side of the Olympic Peninsula; the **information center in Forks** provides visitor information for the west side. Summer information centers are also located at **Storm King** on Lake Crescent and at **Kalaloch** on the ocean coast. The **Hurricane Ridge Visitor Center** is located in scenic subalpine meadowlands 17 miles south of Port Angeles. It is open from late April to the end of October and on weekends and holidays in winter and features an information desk, exhibits, and publications.

A free park newspaper, *The Bugler,* is distributed at visitor centers and entrance stations; it lists services, seasonal schedules, campfire talks, and guided nature walks. There is also a Junior Ranger program for young people. Schedules for interpretive programs are posted at visitor centers and campgrounds. For information, write to Superintendent, Olympic National Park, 600 Park Avenue, Port Angeles, WA 98362, phone (360) 565-3130, or visit the park's web site at www.nps.gov/olym/.

For information on **Olympic National Forest,** call (360) 956-2402, or visit the forest's web site at www.fs.fed.us/r6/olympic/.

For information on the **Olympic Coast National Marine Sanctuary,** call (360) 457-6622 or visit the sanctuary's web site at www.ocnms.nos.noaa.gov/.

Information on the peninsula's national wildlife refuges is available by calling the **Washington Maritime Wildlife Refuge Complex** at (360) 457-8451.

Seasons and Fees

Olympic National Park is open all year. Some roads and facilities are closed in winter. The summer season begins around Memorial Day and extends through Labor Day. Entrance fees are collected from May through September, and at some locations in winter. Campground fees and wilderness use fees are also required. Call the park at (360) 565-3130 for current fee information.

Transportation

Ferries operate regularly across Puget Sound from Seattle, Edmonds, and Keystone. Schedules are available from Washington State Ferries at

1-800-843-3779 or on the web at www.wsdot.wa.gov/ferries/. Ferry service is also available between Victoria, British Columbia, Canada, and Port Angeles. Schedules are available from Black Ball Transport, Inc., at (360) 457-4491.

Sightseeing by Car

U.S. 101 is the main scenic route around the peninsula, with numerous spur roads providing access into the interior and coastal areas of the park. No roads pass through the rugged heart of the Olympics. Temperate rain forests may be seen along the Hoh, Queets, and Quinault valley roads; old-growth Douglas-fir forests are found along the Sol Duc and Elwha roads. The road to Hurricane Ridge offers splendid views from lowland forest to subalpine meadows. Hurricane Ridge offers stunning views into the wilderness heart of the Olympics. Forest roads also access the Deer Park, Dosewallips, and Staircase areas of the park during the summer. The Pacific coastal strip is accessible from U.S. 101 at Kalaloch or by spur roads to Oil City at the mouth of Hoh, La Push, and Rialto Beach.

Camping and Lodges

The park has 16 established campgrounds. Most consist of individual sites with tables and fireplaces. Water and toilet facilities are nearby. No showers, laundry, or utility connections are provided in park campgrounds; a few have dump stations. Additional campgrounds are located in Olympic National Forest. Overnight accommodations in the park are available at Kalaloch Lodge, (360) 962-2271; Lake Crescent Lodge, (360) 928-3211; Log Cabin Resort, (360) 928-3325; and Sol Duc Hot Springs Resort, (360) 327-3583. Kalaloch Lodge is open year-round; the others operate only in summer. Lake Quinault Lodge in Olympic National Forest at (360) 288-2900 is also open year-round. Meals and accommodations are also available at numerous locations outside the park.

The Olympic Park Institute, an environmental education facility, is housed at historic Rosemary Inn on Lake Crescent. It offers a variety of field seminars and Elderhostel programs for adults as well as environmental education programs for youth and mixed-generation programs. The institute can be reached at (360) 928-3720 or on the web at www.yni.org/opi/.

Hiking

Olympic is a wilderness park, best experienced by leaving roads and cars behind and venturing out on park trails. Short, easy loop trails are located close to most campgrounds and visitor centers. The lower Hurricane Hill trail and Meadow Loop trail at Hurricane Ridge, and the first one-half mile of the Marymere Falls trail at Lake Crescent, are wheelchair accessible with assistance. The Madison Falls trail on the Elwha, the Hoh Visitor Center loop trail, the Moments in Time nature trail on Lake Crescent, and the Maple Glades nature trail at Quinault are fully accessible.

More than 500 miles of backcountry trails provide access to the interior mountains of the park. Backcountry rangers are stationed throughout the park in summer months. Some trails are suitable for horseback riding as well as hiking, but pets, bikes, and weapons are prohibited on all park trails. Hikers should be prepared for inclement weather at any time of year; stoves, tents, and warm clothing are advised. Permits are required for all overnight backpackers and climbing parties. Quotas and reservations are in effect for some heavily used wilderness areas. Wilderness use fees are required for overnight trips. Hikers should contact the Wilderness Information Center at (360) 565-3100 or on the web at www.nps.gov/olym/wic/.

Where to See Wildflowers

With its abundant rainfall and mild maritime climate, the Olympic Peninsula harbors a wide variety of wildflowers. Diverse habitats support more than 1,400 species of vascular plants. After the peninsula's heavy winter rains, the first warm days of spring bring coastal areas and lowland valleys into bloom, and the lowland trails of the park become lined with blossoms.

The wildflower season extends into late summer as cool, north-facing slopes of the high mountains melt free of winter snow. Charles Stewart's *Wildflowers of the Olympics and Cascades* is a handy and easy-to-use trailside identification guide tailored to Northwest flora.

Lowland Forest Wildflower Walks

The **Peabody Creek trail,** beginning at the Olympic National Park Visitor Center in Port Angeles, is an excellent place to sample the spring bloom. From mid-April through May, trailsides are dotted with pioneer violet, trillium, false Solomon's seal, twisted stalk, twinflower, fringe cup, and wild ginger. The **Marymere Falls trail** at Lake Crescent hosts similar wildflower displays, and the nearby **Spruce Railroad trail,** with its gravel slopes and rocky outcrops, adds paintbrush, sedum, and spotted saxifrage as well spirea and chocolate lilies to the lowland display.

In the western rain forest, the **Hall of Mosses trail** at the Hoh Visitor Center meanders through pale sprays of white foamflower, slender boykina, and thick beds of oxalis as well as pioneer violet, beadruby, springbeauty, and trillium.

Mountain Forest Wildflower Walks

In late May and June, the wildflowers of the mountain forests are at their peak. The **Hurricane Ridge** and **Deer Park** roads pass through this zone, and trails at **Heart of the Hills, Elwha, Stair-**

case, Dosewallips, Sol Duc, and other locations provide access to the montane forest. Pipsissewa, pyrola, bunchberry, and delicate fairy slippers bloom in this shady realm. These mid-elevation forests are also home to shade-loving plants that lack green chlorophyll: pinedrops, pinesaps, and ghostly Indian pipes.

Subalpine Meadow Wildflower Walks

Beginning in June, snow on the subalpine meadows begins to melt. The **Hurricane Hill** and **Deer Ridge trails** take hikers through carpets of mountain wildflowers. Smooth douglasia and spreading phlox are among the first to bloom on south-facing rocky slopes, and glacier lilies follow the melting edges of snowbanks. Colorful gardens of subalpine buttercup, Sitka valerian, pale larkspur, arnica, hook violet, paintbrush, and cow parsnip cover the hillsides. In the **Hurricane Ridge** area and along the Hurricane Ridge Road, monkey flowers and marsh marigolds crowd snowmelt streams, and purplish-pink Jeffrey's shooting stars add splashes of color to moist swales. Other trails from which to sample the subalpine gardens in early summer include the **Royal Basin** and **Boulder Lake trails** in the park, and the **Mount Townsend** and **Marmot Pass trails** in Olympic National Forest.

Alpine Wildflower Walks

The **Elk Mountain** and **Lillian Ridge trails** from Obstruction Point

venture past the tree line into the Arctic-alpine zone. The windswept ridges and shallow soils of this zone limit plant growth to low-growing and "cushion" plants. Here phlox and douglasia are joined by western and rock-loving anemone, sweet vetch, phacelia, and Webster's senecio (an Olympic endemic). Pale agoseris, woolly pussytoes, small-flowered penstemon, and several saxifrages also inhabit this airy realm.

Park naturalists lead nature walks at several locations throughout the summer season. Check for scheduled walks at park visitor centers.

Where to See Old-Growth Forests

Olympic National Park harbors some of the finest examples of temperate old-growth forests remaining in the Pacific Northwest. Lowland forests can be visited any time of year, and several short, easy loop trails provide access to some of the park's most magnificent forest stands. Robert L. Wood's *Olympic Mountains Trail Guide* provides detailed descriptions of these short forest walks.

Ancient Forests of the Northern and Eastern Olympics

An excellent introduction to the old-growth Douglas-fir and western hemlock forests of Olympic National Park is the **Ancient Groves nature trail** just off the Sol Duc River Road. This half-mile loop trail meanders through an ancient stand of fir and hemlock. The largest trees, with thick, deeply furrowed, reddish-brown bark, are Douglas-firs; smaller trees, with ridged, somewhat scaly, gray-brown bark, are western hemlocks. The Douglas-fir requires a disturbance such as a wildfire or windstorm to create openings in the forest before its seeds can germinate; this stand may have seeded in after wildfires burned this area 300 to 500 years ago. Western hemlock can thrive in the shade of the older fir forest, and if left undisturbed, hemlock will eventually dominate the stand. Old-growth features such as standing dead trees, downed logs, and an open, multistoried canopy can be seen throughout the forest. Because the Sol Duc swings west and drains to the Pacific, the valley draws moist coastal air. This results in a rain forest influence, which is evident in the presence of Sitka spruce and nurse logs.

The **Sol Duc River trail** at the end of the Sol Duc Road explores miles of this valley forest. For an intimate glimpse of one of the Northwest's finest ancient forests, walk the first mile to Sol Duc Falls, or make a loop using the **Lovers Lane trail.**

The **Marymere Falls** and **Moments in Time nature trails** at Lake Crescent explore a stunning grove of ancient grand fir, Douglas-fir, and western redcedar. Most of the Peninsula's lowland stands containing old-growth grand fir were logged long ago. Less than 600 feet above sea level, the Barnes Point forest

is among the few of its kind remaining. The bark of grand fir is grayer and less deeply textured than Douglas-fir. Western redcedar's thin reddish bark is fibrous and somewhat stringy. The trails join and continue to picturesque Marymere Falls or extend into montane forests up Barnes Creek. U.S. 101 passes through this forest in the vicinity of the Storm King Ranger Station, one of the few old-growth stands visible from the highway.

Farther east, the **West Elwha trail** follows the Elwha River downstream from Altaire campground through a mixed old-growth forest of Douglas-fir and western hemlock. The Humes Ranch loop on the Elwha River trail passes beneath stately old Douglas-firs and hemlocks. Some of the larger trees still bear blazes cut from 1889 to 1890 by the Press party.

At Heart o' the Hills on the Hurricane Ridge Road, the **Heart of the Forest trail** passes through second growth and enters an old-growth forest a short distance in from the trail head. The largest of the Douglas-firs and redcedars here approach eight feet in diameter.

The first mile of the **Dungeness River trail** in Olympic National Forest traverses a handsome stand of old-growth Douglas-fir growing on the narrow floodplain along the river. The bark of many of the trees growing in this rainshadow valley has been charred by frequent wildfires; growth beneath them is spare, and the ground is covered with moss.

At Staircase, in the far southeast corner of the park, rainfall is much more plentiful. The mile-long **Shady Lane trail** follows the Skokomish River beneath groves of bigleaf maples and through stunning stands of Douglas-fir, western hemlock, and western redcedar. Many of the larger trees approach diameters of 8 feet and exceed 250 feet in height. The nearby **Staircase Rapids trail** leads upriver through an equally impressive mixed forest of hemlock, cedar, and Douglas-fir.

Rain Forests of the West-Side Valleys

One of the loveliest walks through Olympic's temperate rain forest is along the **Hall of Mosses nature trail** at the Hoh Visitor Center. Starting in a lush riparian area alongside a spring-fed stream, the loop trail ascends a river terrace beneath massive, mossy Douglas-firs, then meanders through a forest of large Sitka spruces and western hemlocks. The spruces have wide-spreading, buttressed, cylindrical trunks and gray scaly bark; some approach 300 feet in height. The forest floor is carpeted with oxalis, mosses, and ferns. The "Hall of Mosses" itself is an enchanting grove of bigleaf maples draped with hanging mats of selaginella clubmoss, mosses, and licorice ferns. Nurse logs, colonnades, and other signature characteristics of the temperate rain forest can be found along this gentle, three-quarter-mile-long trail.

The **Quinault Loop trail,** located on forest service land on the south shore of Quinault Lake, is a network of three trails. The loop trail traverses a grove of huge western redcedars as well as impressive Sitka spruces and western hemlocks. The **Rain Forest nature trail** which joins it ascends through the Big Tree Grove, a magnificent stand of 6 to 8-foot-diameter Douglas-firs that tower more than 275 feet in height. This stand may have seeded in on the ashes of massive fires that burned through the area around 1508.

Park roads up the Hoh and Quinault river valleys, and the main trails up the Bogachiel, Hoh, Queets, North Fork, and East Fork Quinault valleys, also provide access to outstanding examples of temperate rain forest; they remain open year-round.

The Olympic Peninsula hosts several trees of world-record or near-record size. Some are easy to see; some require extended hikes. The trips are always worthwhile. Outstanding specimens include the 14-foot-diameter and 212-foot-high "Queets Fir," located about two and a half miles up the **Queets River trail.** A record-sized Sitka spruce, 18 feet in diameter and just under 200 feet in height, is found on the south shore of Quinault Lake in Olympic National Forest; an equally impressive giant is found on the **South Fork Hoh trail.** A champion western hemlock, measuring 9 feet in diameter and 174 feet in height, is found up the **East**

Fork Quinault trail, 2 miles past the Enchanted Valley chalet in the heart of the mountains. The record western redcedar, measuring more than 20 feet in diameter and 159 feet in height, stands on the north shore of Lake Quinault; a slightly smaller redcedar stands in the park near Kalaloch. The record subalpine fir and Alaska cedar are found in the interior mountains. Check with nearby ranger stations for locations, but bear in mind that large trees can fall or lose their tops, and records often flit from state to state. It is the wholeness and complexity of the community that make Olympic's ancient forests so special, and these characteristics can be savored in any grove of old-growth forest on the peninsula.

Where to See Wildlife

Mammals

Mammals thrive in every corner of Olympic National Park, from rocky shorelines to mountain meadows. Most are small, and some are active only at night. Even large mammals are often difficult to spot in the dense forests of the park. The best times for catching glimpses of Olympic's larger mammals are during the early hours of daylight and at dusk. Many animals retreat to forest shade and rest during the heat of the day.

Columbia blacktail deer are the most commonly encountered large mammals in the park, and visitors staying any length of time are almost sure to see them. Deer are often seen feeding along park roadsides and at the edges of forest clearings. Typically, does are seen with one or two young. Many deer summer in Big Meadow at Hurricane Ridge. Because of the large number of visitors to the ridge, these deer seem to have lost their fear of humans and sometimes approach them quite closely. Remember it is unlawful, and ultimately harmful, to feed any wildlife in the park. Deer are also common at Deer Park and at most park campgrounds.

Roosevelt elk are a bit more elusive. They are frequently spotted in valley forests along the Hoh, Queets, or Quinault roads. Look for their blond rump patches among the trees. Bunch Field, just below the forks of the Quinault River, is an excellent spot for morning or evening viewing, and elk are often seen around the Hoh campground. All rain forest trails pass through prime elk habitat. Look for their large split tracks in wet spots, and watch for browsed ferns and shrubs.

Elk are also seen along the lower Elwha River in the park, especially at Anderson field and Humes Ranch along the Elwha trail. Roughly half of the park's elk migrate to the high meadows of the interior mountains in summer. Good places to watch for them are Heart Lake and High Divide above the Sol Duc Valley, and Low Divide and Skyline Divide in the Quinault area.

When rutting season begins in September, listen for the shrill calls of the bulls throughout the high country and in valley forests.

In summer, when mountain berries ripen, **black bears** also head to the high meadows to put on fat for their winter hibernations. Bears appear around Hurricane Hill in early summer some years, nibbling new plants and sampling the bark of subalpine firs. Later in the summer, look for them in the meadow country around Hayden and Anderson passes, the Marmot Lake area, Enchanted Valley, and Low Divide, as well as the high country around Seven Lakes Basin and Heart Lake.

Marmots also wake early in the season and can be seen from the Hurricane Hill trail or the trail through Big Meadow behind the Hurricane Ridge Visitor Center. Look for them also at Deer Park and Obstruction Point. Subalpine meadows at Hurricane Ridge and Deer Park also support healthy populations of **snowshoe hares** and the endemic **short-tailed weasels** that hunt them. Another endemic subspecies, the **Olympic chipmunk,** keeps busy around the forested edges of these meadows, and lots of smaller mammals scurry around beneath the brush.

Although **cougar, bobcat,** and **coyote** are known to hunt the high meadows as well as the lower forests, all are fairly reclusive and are seldom seen by visitors. **River otters** and **raccoons,** however, are quite common, especially along valley streams. Look for their five-toed tracks in the soft mud of stream banks, particularly along the coastal strip.

Sea Mammals

California gray whales often feed close to shore on their northern migrations in spring. Beginning in March and April, Cape Flattery, Cape Alava, La Push, Ruby Beach, and Kalaloch are good places to watch for them. A few whales sometimes linger in coastal waters or in the Strait of Juan de Fuca for the summer.

Harbor seals are abundant along the park's coastal strip any time of year. Look for them hauled out on offshore rocks at low tide or surfacing in offshore kelp beds. Harbor seal pups also haul out on the Dosewallips, Duckabush, and Hamma Hamma estuaries along Hood Canal, and at the end of Dungenesss Spit in the strait. Seal pups—or any sea mammals—found along park beaches should be left alone; report such sightings to the nearest ranger station.

California and **Steller's sea lions** are a bit harder to see from shore. Look for them in early summer and fall on their haul-outs on Tatoosh Island off Cape Flattery and on the Bodelteh Islands off Cape Alava. **Sea otters** love to float among kelp beds through much of the day, feeding actively in mornings and late afternoons. They are best seen from atop coastal headlands where nearby kelp beds are visible. Good viewing spots are at Cape Flattery, Cape Alava, Sand Point, and Cape Johnson. **River otters** are often confused with sea otters, especially when they are swimming in the surf. River otters are smaller than sea otters; unlike sea otters, they are quite agile on land, and they never float on their backs.

Birds

Marine birds and shorebirds can be seen anywhere along the park's coastal strip as well as in the Strait of Juan de Fuca and parts of Hood Canal. Shorebirds such as **sandpipers, snipes, dunlins,** and **semipalmated plovers** frequent sandy beaches at Shi Shi, Yellow Banks, Rialto Beach, and the beaches from Ruby Beach south to Kalaloch. Dungeness Spit National Wildlife Refuge, near Sequim, is also an excellent place for viewing shorebirds. Seabirds of the rocky shore, such as **oystercatchers, cormorants, scoters, surfbirds,** and **murres,** can be seen by taking short hikes to Cape Flattery, Cape Alava, Cape Johnson, Second and Third beaches near La Push, and Hoh Head. **Bald eagles** and **ospreys** are seen almost anywhere along the park's coastal strip.

Beginning in late March and extending through April, one of the largest raptor migrations in North America occurs along the Olympic coast. Thousands of hawks: **red-tailed, sharp-shinned, Cooper's,** and **goshawks** as well as **kestrels, eagles, ospreys, turkey**

vultures, and other raptors stream north on their spring migrations. At the northern tip of the peninsula, they ride the thermals above Cape Flattery, gaining the altitude needed to cross the strait to Vancouver Island. The summit of Bahokus Peak, just west of Neah Bay on the Makah Indian Reservation, is an excellent place to view this spectacle. On warm days during the peak of the migration, as many as 600 or more raptors may make the crossing.

More than 200 species of resident and migratory birds are found in Olympic National Park. Many inhabit a range of habitats over different seasons, and songbirds may be encountered anywhere in the park. In the forests of low and middle elevations, birds are more often heard than seen, and many birders rely on songs for identification. During the breeding season, which begins in March, most lowland forest trails and roadsides ring with birdsong.

Visitors to the Hoh, Queets, or Quinault rain forests can expect to encounter **yellow warblers, chestnut-backed chickadees,** and **Vaux's swifts,** among other birds. **Winter wrens** sing among the ferns and brush, and **golden-crowned kinglets** sweep through the canopy. Look for **warblers, chickadees,** and **thrushes** in the deciduous growth along rivers and streams.

In the drier Douglas-fir and hemlock forests of the northern and eastern Olympics — the Elwha, Dungeness, Dosewallips, Duckabush, and Skokomish valleys — watch and listen for **Townsend's warblers, evening grosbeaks, olive-sided flycatchers, and varied** and **Swainson's thrushes,** as well as **chickadees, warblers, northern flickers, and brown creepers.**

Birds are less abundant in the cooler montane forests of higher elevations. In the silver fir and hemlock forests of the western slopes, look for **varied thrushes, winter wrens, dark-eyed juncos, and** both **Steller's** and **gray jays.** Subalpine fir forests of the eastern high country support breeding populations of **olive-sided flycatchers, ruby-crowned kinglets, pine grosbeaks, Townsend's solitaires,** and **blue grouse.**

The open alpine areas of the park are accessible along the road to Obstruction Point and by trails leaving from the Obstruction Point parking area and the Hurricane Hill trail. Alpine areas are breeding territories for **horned larks, American pipits,** and **dark-eyed juncos. Rosy finches** and occasional **golden eagles** may also be seen here. In late summer and fall, south-migrating raptors hunt the high meadows and ridges.

A final avian feast takes place in the high meadows each fall as huckleberries and mountain ash berries ripen. **Band-tailed pigeons, robins, thrushes,** and **sparrows** come in flocks to feed. As the first fall snows dust the high country and heavy rains dampen the lower forests, many of the park's songbirds begin their migrations south.

SPECIES CHECKLISTS: PLANTS AND ANIMALS OF OLYMPIC NATIONAL PARK

SELECTED WILDFLOWERS

Common Yarrow	*Achillea millifolium*
California Vanillaleaf	*Achlys californica*
Trail Plant	*Adenocaulon bicolor*
Pale Agoseris	*Agoseris glauca*
Scalloped Onion	*Allium cernuum*
Nodding Onion	*Allium crenulatum*
Pearly-everlasting	*Anaphalis margaritacea*
Pacific Anemone	*Anemone multifida*
Western Pasqueflower	*Anemone occidentalis*
Sitka Columbine	*Aquilegia formosa*
Broad-leaf Arnica	*Arnica latifolia*
Wild Ginger	*Asarum caudatum*
Olympic Mountain Aster	*Aster paucicapitatus*
Elkslip MarshMarigold	*Caltha leptosepala*
Fairy Slipper	*Calypso bulbosa*
Piper's Bellflower	*Campanula piperi*
Bluebells-of-Scotland	*Campanula rotundifolia*
Merten's Mountain-heather	*Cassiope mertensiana*
Harsh Paintbrush	*Castilleja hispida*
Magenta Paintbrush	*Castilleja parviflora*
Common Pipsissewa	*Chimaphila umbellata*
Indian Thistle	*Cirsium edule*
Lance-leaved Springbeauty	*Claytonia lanceolata*
Siberian Springbeauty	*Claytonia sibirica*
Queen's Cup	*Clintonia uniflora*
Unalaska Bunchberry	*Cornus unalaschkensis*
Rockslide Larkspur	*Delphinium glareosum*
Pale Larkspur	*Delphinium glaucum*
Pacific Bleeding Heart	*Dicentra formosa*
Jeffrey's Shooting Star	*Dodecatheon jeffreyi*
Smooth Douglasia	*Douglasia laevigata*
Elmera	*Elmera racemosa*
Alpine Willow-herb	*Epilobium alpinum*
Fireweed	*Epilobium angustifolium*
Wandering Fleabane	*Erigeron peregrinus*
Woolly Sunflower	*Eriophyllum lanatum*
Western Wallflower	*Erysimum capitatum*
Glacier Lily	*Erythronium grandiflorum*
Avalanche Lily	*Erythronium montanum*

Mission Bells	*Fritillaria lanceolata*
Fragrant Bedstraw	*Galium triflorum*
Old man's whiskers	*Geum triflorum*
Cow Parsnip	*Heracleum lanatum*
Columbia Lily	*Lilium columbianum*
Twinflower	*Linnaea borealis*
Martindale's Lomatium	*Lomatium martindalei*
Orange Honeysuckle	*Lonicera ciliosa*
Partridgefoot	*Luetkea pectinata*
Broadleaf Lupine	*Lupinus latifolius*
Lyall's Lupine	*Lupinus lyallii*
Yellow Skunk Cabbage	*Lysichiton americanum*
Beadruby	*Maianthemum dilatatum*
Large False Solomon's Seal	*Maianthemum racemosum*
Great Purple Monkey-flower	*Mimulus lewisii*
Large Mountain Monkey-flower	*Mimulus tilingi*
Woodnymph	*Moneses uniflora*
Pinesap	*Monotropa hypopithys*
Indian-pipe	*Monotropa uniflora*
Mountain Owl-clover	*Orthocarpus imbricatus*
Oregon Oxalis	*Oxalis oregana*
Elephant's Head	*Pedicularis groenlandica*
Sickletop Lousewort	*Pedicularis racemosa*
Davidson's Penstemon	*Penstemon davidsonii*
Small-flowered Penstemon	*Penstemon procerus*
Olympic Rockmat	*Petrophytum hendersonii*
Silky Phacelia	*Phacelia sericea*
Spreading Phlox	*Phlox diffusa*
Red Mountain-heather	*Phyllodoce empetriformis*
Greater Butterwort	*Pinguicula macroceras*
Dilated Bog Orchid	*Platanthera dilatata*
Showy Polemonium	*Polemonium pulcherrimum*
American Bistort	*Polygonum bistortoides*
Fan-leaf Cinquefoil	*Potentilla flabellifolia*
Self-heal	*Prunella vulgaris*
Pinedrops	*Pterospora andromedea*
Western Pasqueflower	*Pulsatilla occidentalis*
Common Pink Pyrola	*Pyrola asarifolia*
Eshscholtz's Buttercup	*Ranunculus eschscholtzii*
Dwarf Bramble	*Rubus lasiococcus*
Tufted Saxifrage	*Saxifraga cespitosa*
Broad-leaved Sedum	*Sedum spathulifolium*
Olympic Mountain Groundsel	*Senecio neowebsteri*

Parry's Silene   *Silene parryi*
Subalpine Spirea   *Spiraea densiflora*
Clasping-leaved Twisted-stalk   *Streptopus amplexifolius*
Fringecup   *Tellima grandiflora*
Trefoil Foamflower   *Tiarella trifoliata*
Youth-on-age   *Tolmiea menziesii*
American Starflower   *Trientalis borealis*
Large White Trillium   *Trillium ovatum*
Sitka Valerian   *Valeriana sitchensis*
Green False Hellebore   *Veratrum viride*
Cusick's Speedwell   *Veronica cusickii*
American Vetch   *Vicia americana*
Hook Violet   *Viola adunca*
Flett's Violet   *Viola flettii*
Pioneer Violet   *Viola glabella*
Beargrass   *Xerophyllum tenax*

SELECTED TREES
Pacific Silver Fir   *Abies amabilis*
Grand Fir   *Abies grandis*
Subalpine Fir   *Abies lasiocarpa*
Vine Maple   *Acer circinatum*
Douglas Maple   *Acer glabrum douglasii*
Bigleaf Maple   *Acer macrophyllum*
Speckled Alder   *Alnus incana*
Red Alder   *Alnus ruba*
Sitka Alder   *Alnus sinuata*
Pacific Madrona   *Arbutus menziesii*
Alaska Yellow Cedar   *Chamaecyparis nootkakatensis*
Pacific Dogwood   *Cornus nuttallii*
Douglas' Hawthorn   *Crataegus douglasii*
Oregon Ash   *Fraxinus latifolia*
Common Juniper   *Juniperus communis*
Rocky Mountain Juniper   *Juniperus scopulorum*
Engelmann Spruce   *Picea engelmanni*
Sitka Spruce   *Picea sitchensis*
Whitebark Pine   *Pinus albicaulis*
Lodgepole Pine   *Pinus contorta*
Quaking Aspen   *Populus tremuloides*
Black Cottonwood   *Populus trichocarpa*
Bitter Cherry   *Prunus emarginata*
Common Chokecherry   *Prunus virginiana*

Douglas-fir	*Pseudotsuga menziesii*
Western Crabapple	*Pyrus fusca*
Garry Oak	*Quercus garryana*
Cascara Buckthorn	*Rhamnus purshiana*
Pacific Willow	*Salix lasiandra*
Pacific Yew	*Taxus brevifolia*
Western Redcedar	*Thuja plicata*
Western Hemlock	*Tsuga heterophylla*
Mountain Hemlock	*Tsuga mertensiana*

SELECTED SHRUBS

Western Serviceberry	*Amelanchier alnifolia*
Bristly Manzanita	*Arctostaphylos columbiana*
Kinnikinnick	*Arctostaphylos uva-ursi*
Goatsbeard	*Aruncus sylvester*
Oregon Grape	*Berberis nervosa*
Redstem Ceanothus	*Ceanothus sanguineus*
Creek Dogwood	*Cornus stolonifera*
Salal	*Gaultheria shallon*
Ocean-spray	*Holodiscus discolor*
Black Twinberry	*Lonicera involucrata*
Fool's Huckleberry	*Menziesia ferruginea*
Indian Plum	*Oemleria cerasiformus*
Devil's Club	*Oplopanax horridus*
Ninebark	*Physocarpus capitatus*
White Rhododendron	*Rhododendron albiflorum*
Pacific Rhododendron	*Rhododendron macrophyllum*
Swamp Gooseberry	*Ribes lacustre*
Red-flowering Currant	*Ribes sanguineum*
Baldhip Rose	*Rosa gymnocarpa*
Nootka Rose	*Rosa nutkana*
Western Blackcap	*Rubus leucodermis*
Western Thimbleberry	*Rubus parviflorus*
Salmonberry	*Rubus spectabilis*
Hooker Willow	*Salix hookeriana*
Scouler Willow	*Salix scouleriana*
Sitka Willow	*Salix sitchensis*
Blue Elderberry	*Sambucus cerulea*
Coast Red Elderberry	*Sambucus racemosa*
Sitka Mountain-ash	*Sorbus sitchensis*
Douglas' Spirea	*Spiraea douglasii*
Dwarf Huckleberry	*Vaccinium cespitosum*
Blue-leaf Huckleberry	*Vaccinium deliciosum*

Oval-leaf Huckleberry	*Vaccinium ovalifolium*
Evergreen Huckleberry	*Vaccinium ovatum*
Red Huckleberry	*Vaccinium parvifolium*

SELECTED FERNS

Maidenhair Fern	*Adiantum pedatum*
Lady Fern	*Athyrium filix-femina*
Deer Fern	*Blechnum spicant*
Spreading Wood Fern	*Dryopteris expansa*
Oak Fern	*Gymnocarpium dryopteris*
Licorice Fern	*Polypodium glycyrrhiza*
Sword Fern	*Polystichum munitum*
Bracken Fern	*Pteridium aquilinum*

MAMMALS

Masked Shrew	*Sorex cinereus*
Trowbridge Shrew	*Sorex trowbridgei*
Vagrant Shrew	*Sorex vagrans*
Dusky Shrew	*Sorex monticolus*
Pacific Water Shrew	*Sorex bendirii*
Northern Water Shrew	*Sorex palustris*
Shrew Mole	*Neurotrichus gibbsii*
Townsend's Mole	*Scapanus townsendii*
Snow Mole	*Scapanus townsendii olympicus*
Coast Mole	*Scapanus orarius*
Little Brown Myotis	*Myotis lucifugus*
Yuma Myotis	*Myotis yumanensis*
Keen Myotis	*Myotis keeni*
Long-eared Myotis	*Myotis evotis*
California Myotis	*Myotis californicus*
Long-legged Myotis	*Myotis volans*
Western Big-eared Bat	*Plecotus townsendi*
Showshoe Hare	*Lepus americanus*
Aplodontia (Mountain Beaver)	*Aplodontia rufa*
Douglas Squirrel	*Tamiasciurus douglasii*
Northern Flying Squirrel	*Glaucomys sabrinus*
Townsend's Chipmunk	*Eutamias townsendi*
Olympic Chipmunk	*Eutamias amoenus caurinus*
Beaver	*Castor canadensis*
Muskrat	*Ondatra zibethicus*
Olympic Marmot	*Marmota olympus*
Northern Pocket Gopher	*Thomomys talpoides*

Mazama Pocket Gopher	*Thomomys mazama melanops*
Deer Mouse	*Peromyscus maniculatus*
Bushy-tailed Woodrat	*Neotoma cinerea*
Mountain Heather Vole	*Phenacomy intermedius*
Southern Redback Vole	*Clethrionomys gapperi*
Townsend Vole	*Microtus townsendi*
Long-tailed Vole	*Microtus longicaudus*
Oregon Vole	*Microtus oregoni*
Pacific Jumping Mouse	*Zapus princeps*
Coyote	*Canis latrans*
Wolf	*Canis lupus*
Red Fox	*Vulpes fulva*
Black Bear	*Ursus americanus*
Raccoon	*Procyon lotor*
Marten	*Martes americana*
Fisher	*Martes pennanti*
Short-tailed Weasel	*Mustela erminea*
Long-tailed Weasel	*Mustela frenata*
Mink	*Mustela vison*
River Otter	*Lutra canadensis*
Spotted Skunk	*Spilogale putorius*
Striped Skunk	*Mephitis mephitis*
Cougar	*Felis concolor*
Bobcat	*Lynx rufus*
Roosevelt Elk	*Cervus elaphus roosevelti*
Columbia Blacktail Deer	*Odocoileus hemionus columbianus*

SELECTED MARINE MAMMALS

Sea Otter	*Enhydra lutris*
Harbor Seal	*Phoca vitulina*
Steller's Sea Lion	*Eumetopias jubatus*
California Sea Lion	*Zalophus californianus*
Northern Elephant Seal	*Mirounga angustirostris*
Northern Fur Seal	*Callorhinus ursinus*
Gray Whale	*Eschrichtius robustus*
Minke Whale	*Balaenoptera acutorostrata*
Humpback Whale	*Megaptera novaeangliae*
Harbor Porpoise	*Phocoena*
Killer Whale	*Orcinus orca*
Dall's Porpoise	*Phocoenoides dalli*
Pacific White-sided Dolphin	*Lagenorhynchus obliquidens*

SELECTED BIRDS

Red-throated Loon
Pacific Loon
Common Loon
Pied-billed Grebe
Horned Grebe
Red-necked Grebe
Eared Grebe
Western Grebe
Brown Pelican
Double-crested Cormorant
Brandt's Cormorant
Pelagic Cormorant
Black-footed Albatross
Northern Fulmar
Pink-footed Shearwater
Flesh-footed Shearwater
Sooty Shearwater
Fork-tailed Storm-petrel
Leach's Storm-petrel
American Bittern
Great Blue Heron
Great Egret
Tundra Swan
Trumpeter Swan
Great White-fronted Goose
Brant
Canada Goose
Wood Duck
Green-winged Teal
Mallard
Northern Pintail
Blue-winged Teal
Cinnamon Teal
Northern Shoveler
Gadwall
American Wigeon
Eurasian Wigeon
Canvasback
Ring-necked Duck
Greater Scaup
Lesser Scaup

Harlequin Duck
Oldsquaw
Black Scoter
Surf Scoter
White-winged Scoter
Common Goldeneye
Barrow's Goldeneye
Bufflehead
Hooded Merganser
Common Merganser
Red-breasted Merganser
Ruddy Duck
Turkey Vulture
Osprey
Northern Harrier
Sharp-shinned Hawk
Cooper's Hawk
Northern Goshawk
Red-tailed Hawk
Golden Eagle
Bald Eagle
American Kestrel
Merlin
Peregrine Falcon
Gyrfalcon
Ring-necked Pheasant
Blue Grouse
Ruffed Grouse
California Quail
Virginia Rail
Sora
American Coot
Black-bellied Plover
Lesser Golden Plover
Semipalmated Plover
Killdeer
Black Oystercatcher
Greater Yellowlegs
Lesser Yellowlegs
Willet
Spotted Sandpiper

Whimbrel
Long-billed Curlew
Marbled Godwit
Ruddy Turnstone
Black Turnstone
Surfbird
Red Knot
Sanderling
Western Sandpiper
Least Sandpiper
Rock Sandpiper
Dunlin
Short-billed Dowitcher
Long-billed Dowitcher
Common Snipe
Red-necked Phalarope
Red Phalarope
Bonaparte's Gull
Heermann's Gull
Mew Gull
Ring-billed Gull
California Gull
Thayer's Gull
Western Gull
Glaucous-winged Gull
Black-legged Kittiwake
Caspian Tern
Common Murre
Pigeon Guillemot
Marbled Murrelet
Ancient Murrelet
Cassin's Auklet
Rhinoceros Auklet
Tufted Puffin
Band-tailed Pigeon
Mourning Dove
Rock Dove
Common Barn Owl
Western Screech-Owl
Great Horned Owl
Snowy Owl
Northern Pygmy-Owl
Spotted Owl
Barred Owl

Short-eared Owl
Northern Saw-whet Owl
Common Nighthawk
Black Swift
Vaux's Swift
Anna's Hummingbird
Rufous Hummingbird
Belted Kingfisher
Red-breasted Sapsucker
Downy Woodpecker
Hairy Woodpecker
Northern Flicker
Pileated Woodpecker
Olive-sided Flycatcher
Western Wood-pewee
Willow Flycatcher
Hammond's Flycatcher
Pacific Slope Flycatcher
Horned Lark
Purple Martin
Tree Swallow
Violet-green Swallow
Cliff Swallow
Barn Swallow
Gray Jay
Steller's Jay
Clark's Nutcracker
American Crow
Common Raven
Black-capped Chickadee
Chestnut-backed Chickadee
Bushtit
Red-breasted Nuthatch
Brown Creeper
Bewick's Wren
House Wren
Winter Wren
Marsh Wren
American Dipper
Golden-crowned Kinglet
Ruby-crowned Kinglet
Townsend's Solitaire
Swainson's Thrush
Hermit Thrush

American Robin
Varied Thrush
American Pipit
Cedar Waxwing
Northern Shrike
European Starling
Hutton's Vireo
Warbling Vireo
Orange-crowned Warbler
Yellow Warbler
Yellow-rumped Warbler
Black-throated Gray Warbler
Townsend's Warbler
Hermit Warbler
MacGillivray's Warbler
Common Yellowthroat
Wilson's Warbler
Western Tanager
Black-headed Grosbeak
Rufous-sided Towhee
Savannah Sparrow

Fox Sparrow
Song Sparrow
Lincoln's Sparrow
Golden-crowned Sparrow
White-crowned Sparrow
Dark-eyed Junco
Red-winged Blackbird
Western Meadowlark
Brewer's Blackbird
Brown-headed Cowbird
Rosy Finch
Pine Grosbeak
Purple Finch
House Finch
Red Crossbill
White-winged Crossbill
Pine Siskin
American Goldfinch
Evening Grosbeak
House Sparrow

AMPHIBIANS AND REPTILES

Northwestern Salamander	*Ambystoma gracile*
Long-toed Salamander	*Ambystoma macrodactylum*
Olympic Salamander	*Rhyacotriton olympicus*
Cope's Giant Salamander	*Dicamptodon copei*
Rough-skinned Newt	*Taricha granulosa*
Ensatina Salamander	*Ensatina eschscholtzi*
Western Red-backed Salamander	*Plethodon vehiculum*
Van Dyke's Salamander	*Plethodon vandykei*
Western Toad	*Bufo boreas*
Tailed Frog	*Ascaphus truei*
Red-legged Frog	*Rana aurora*
Pacific Treefrog	*Pseudacris regilla*
Bullfrog	*Rana catesbeiana*
Cascades Frog	*Rana cascadae*
Western Fence Lizard	*Sceloporus occidentalis*
Northern Alligator Lizard	*Gerrhonotus coeruleus*
Rubber Boa	*Charina bottae*
Gopher Snake	*Pituophis melanoleucus*
Common Garter Snake	*Thamnophis sirtalis*
Western Terrestrial Garter Snake	*Thamnophis elegans*
Northwestern Garter Snake	*Thamnophis ordinoides*

SELECTED FISH

Chinook Salmon	*Oncorynchus tshawytscha*
Coho Salmon	*Oncorynchus kisutch*
Sockeye Salmon	*Oncorynchus nerka*
Chum Salmon	*Oncorynchus keta*
Pink Salmon	*Oncorynchus gorbuscha*
Rainbow Trout or Steelhead	*Oncorynchus mykiss*
Cutthroat Trout	*Oncorynchus clarki*
Dolly Varden	*Salvelinus malma*
Bull Trout	*Salvelinus confluentus*
Eastern Brook Trout	*Salvelinus fontinalis*
Pacific Lamprey	*Intosphenus tridentatus*
River Lamprey	*Lampetra ayresi*
Mountain Whitefish	*Prosopium williamsoni*
Pygmy Whitefish	*Prosopium coulteri*
Longfin Smelt	*Spirinchus thaleichthys*
Surf Smelt	*Hypomesus pretiosus*
Olympic Mudminnow	*Novumbra hobbsi*
Northern Squawfish	*Ptychocheilus oregonensis*
Pacific Staghorn Sculpin	*Leptocottus armatus*
Torrent Sculpin	*Cottus rhotheus*
Prickly Sculpin	*Cottus asper*
Threespine Stickleback	*Gasterosteus aculeatus*
Starry Flounder	*Platichthys stellaus*
Pacific Sand Lance	*Ammodytes hexapterus*

SELECTED INTERTIDAL PLANTS

Alaria Kelp	*Alaria marginata*
Coral Leaf Coralline	*Bossiella* spp.
Graceful Coral Coralline	*Corallina vancouveriensis*
Acid Kelp	*Desmarestia ligulata*
Feather Boa Kelp	*Egregia menziesii*
Sea Moss	*Endocladia muricata*
Rockweed	*Fucus* spp.
Rough Strap	*Gigartina harveyana*
Sea Sac	*Halosaccion glandiforme*
Sea Cabbage	*Hedophyllum sessile*
Rainbow Seaweed	*Iridaea cordata*
Split Kelp	*Laminaria setchellii*
Sea Cauliflower	*Leathesia difformis*
Rock Crust	*Lithothamnium philippi*
Perennial Kelp	*Macrocystis integrifolia*

Bull Kelp	*Nereocystis luetkeana*
Sea Brush	*Odonthalia floccosa*
Surfgrass	*Phyllospadix scouleri*
Red Laver (Dulse)	*Porphyra* spp.
Sea Palm	*Postelsia palmaeformis*
Iodine Seaweed	*Prionitis lanceolata*
Black Pine Seaweed	*Rhodomela larix*
Sea Lettuce	*Ulva* spp.
Eelgrass	*Zostera marina*

SELECTED INTERTIDAL ANIMALS

White Cap Limpets	*Acmaea mitra*
Ostrich Plume Hydroid	*Aglaophenia* spp.
Black and White Brittle Sea Star	*Amphipholis pugetana*
Small Brittle Sea Star	*Amphipholis squamata*
Aggregate Sea Anemone	*Anthopleura elegantissima*
Giant Green Sea Anemone	*Anthopleura xanthogrammica*
Cup Coral	*Balanophyllia elegans*
Thatched Acorn Barnacles	*Balanus cariosus*
Common Acorn Barnacles	*Balanus glandula*
Giant Acorn Barnacles	*Balanus nubilus*
Threaded Horn Snail	*Bittium eschrichtii*
Yellow-edged Nudibranch	*Cadlina luteomarginata*
Ghost Shrimp	*Callianassa californiensis*
Blue Top-shelled Snail	*Calliostoma ligatum*
Dungeness Crab	*Cancer magister*
Red Rock Crab	*Cancer productus*
Small Acorn Barnacle	*Chthamalus dalli*
Water-line Isopod	*Cirolana kincaidi*
Heart Cockle Clam	*Clinocardium nuttallii*
Finger Limpet	*Collisella digitalis*
Shield Limpet	*Collisella pelta*
Gum Boot Chiton	*Cryptochiton stelleri*
Sand Dollar	*Dendraster excentricus*
Keyhole Limpet	*Diodora aspera*
Brooding Sea Anemone	*Epiactis prolifera*
White Sea Cucumber	*Eupentacta quinquesemita*
Mottled Sea Star	*Evasterias troschelii*
Hairy Crab	*Hapalogaster mertensii*
Purple Shore Crab	*Hemigrapsus nudus*
Green Shore Crab	*Hemigrapsus oregonensis*
Blood Sea Star	*Henricia leviuscula*

Opalescent Nudibranch	*Hermissenda crassicornis*
Black Chiton	*Katharina tunicata*
Pelagic Goose Barnacle	*Lepas anatifera*
Six-rayed Sea Star	*Leptasterias hexactis*
Pearly Top Shell Snail	*Lirularia lirulata*
Checkered Periwinkle	*Littorina scutulata*
Sitka Periwinkle	*Littorina sitkana*
Vermilion Sea Star	*Mediaster aequalis*
Hairy Chiton	*Mopalia lignosa*
Mossy Chiton	*Mopalia muscosa*
California Blue Mussel	*Mytilus californianus*
Edible Blue Mussel	*Mytilus edulis*
Speckled Limpet	*Notoacmea persona*
Oyster Drill Snail	*Ocenebra japonica*
Purple Olive Snail	*Olivella biplicata*
Daisy Brittle Sea Star	*Ophiopholis aculeata*
Small Beach Hopper	*Orchestia traskiana*
California Beach Hopper	*Orchestoidea californiana*
Decorator Crab	*Oregonia gracilis*
Granular Hermit Crab	*Pagurus granosimanus*
Hairy Hermit Crab	*Pagurus hirsutiusculus*
California Sea Cucumber	*Parastichopus californicus*
Piddock Clam	*Penitella penita*
Green Sea Slug	*Phyllaplysia taylori*
Encrusting Sponge	Phylum *Porifera*
Purple or Ochre Sea Star	*Pisaster ochraceus*
Moon Snail	*Polinices lewisii*
Goose Neck Barnacle	*Pollicipes polymerus*
Native Littleneck Clam	*Protothaca staminea*
Armored Sea Cucumber	*Psolus chitonoides*
Sea Pen	*Ptilosarcus gurneyi*
Graceful Kelp Crab	*Pugettia gracilis*
Kelp Crab	*Pugettia producta*
Sunflower Sea Star	*Pycnopodia helianthoides*
Butter Clams	*Saxidomus giganteus*
Feather Duster Tube Worm	*Schizobranchia insignis*
Spindle Whelk	*Searlesia dira*
Calcareous Tube Worm	*Serpula vermicularis*
Razor Clam	*Siliqua patula*
Sun Sea Star	*Solaster stimpsoni*
Green Sea Urchin	*Strongylocentrotus droebachiensis*
Red Sea Urchin	*Strongylocentrotus franciscanus*
Purple Sea Urchin	*Strongylocentrotus purpuratus*

Red and Green Sea Anemone	*Tealia crassicornis*
Brown Turban Shell Snail	*Tegula brunnea*
Black Turban Shell Snail	*Tegula funebralis*
Helmet Crab	*Telmessus cheiragonus*
Channeled Whelk	*Thais canaliculata*
Wrinkled Whelk	*Thais lamellosa*
Lined Chiton	*Tonicella lineata*
Horse Clam	*Tresus capax*
Orange-spotted Nudibranch	*Triopha carpenteri*

CHAPTER NOTES

Chapter 1 Shuffled Texts of Stone
The best guide to the geology of the Olympic Mountains is Rowland Tabor's *Guide to the Geology of Olympic National Park*, Pacific Northwest Parks and Forests Association, Seattle, 1987. A good overall guide to Pacific Northwest geology is Bates McKee's *Cascadia, the Geologic Evolution of the Pacific Northwest*, McGraw Hill, New York, 1972. A detailed geological map of the Olympics is available from the U.S. Geological Survey, 1978.

History of the geological exploration of the Olympics: C. E. Weaver, *Tertiary Stratigraphy of Western Washington and Northwestern Oregon*, University of Washington Publications in Geology, Vol. 4, 1937, and Tabor, 1987.

Deposition of sedimentary rocks and geological processes: John S. Shelton, *Geology Illustrated*, W. H. Freeman and Co., New York, 1966.

Origin of Olympic rocks: Paul Heller, R. Tabor, J. O'Neil, D. Pevear, M. Shafiqullah, and N. Winslow, "Isotopic Provenance of Paleogene Sandstones from the Accretionary Core of the Olympic Mountains, Washington," *Geological Society of America Bulletin*, Vol. 104, 1992.

Deposition by a single river: Lingly, William S. "Preliminary Observations on Marine Stratigraphic Sequences, Central and Western Olympic Peninsula, Washington," *Washington Geology*, Vol. 23, No. 2, 1995.

Plate tectonics revolution: John McPhee, *Basin and Range*, Farrar, Straus, Giroux, New York, 1981.

Basalt formation, Basin and Range expansion, and subduction: Mark Brandon, and A. Calderwood, "High-pressure Metamorphism and Uplift of the Olympic Subduction Complex," *Geology*, Vol. 18, 1990; also Mark Brandon, and J. Vance, "Tectonic Evolution of the Cenozoic Olympic Subduction Complex, Washington State, as Deducted From Fission Track Ages for Detrital Zircons," *American Journal of Science*, Vol. 292, 1992.

New theory on basalt formation: R. S. Babcock, C. A. Suczek, and D. C. Engebretson, "The Crescent 'Terrane', Olympic Peninsula and Southern Vancouver Island," in R. Lasmanis and E. Cheny, *Regional Geology of Washington State*, Washington State Department of Natural Resources, 1994.

Accretion of Northwest terranes: C. J. Yorath, *Where Terranes Collide*, Orca Publishers, Victoria, 1990.

New dates for core rocks: Brandon and Vance, 1992.

Pre-Ice Age erosion: Tabor, 1987, and Shelton,1966.

Geologic structure of the Olympics: R. W. Tabor and W. M. Cady, *Geological Map of the Olympic Peninsula*, U.S. Geological Survey, 1978.

Continued uplift: Brandon and Calderwood, 1990.

Geology along the Hurricane Ridge Road: R. W. Tabor, "A Tertiary Accreted Terrane: Oceanic Basalt and Sedimentary Rocks in the Olympic Mountains, Washington," *Geological Society of America Centennial Field Guide*, 1987.

Chapter 2 Legacies of Ice
Formation of glaciers: Shelton, 1966.

Dynamics of glaciers: Robert P. Sharp, *Glaciers*, University of Oregon Press, Eugene, 1960.

Origins of ice ages: Brian Skinner and Stephen Porter, *The Dynamic Earth: An Introduction to Physical Geology*, John Wiley and Sons, Inc., 1989.

Glacier carving of the Olympics: Tabor, 1987, and Robert Burns, *The Shape of Puget Sound*, Washington Sea Grant/Puget Sound Books, Seattle, 1985.

Glacier evidence in sea cliffs: Linda Florer, "Quaternary Paleoecology and Stratigraphy of the Sea Cliffs, Western Olympic Peninsula, Washington," *Quaternary Research* 2, 1972.

Continental glaciers in Puget Sound and the Strait of Juan de Fuca: Arthur R. Kruckeberg, *The Natural History of Puget Sound Country*, University of Washington Press, Seattle, 1991; also Don Easterbrook, "Stratigraphy and Chronology of Quaternary Deposits of the Puget Lowland and Olympic Mountains of Washington and Cascade Mountains of Washington and Oregon," *Quaternary Science Reviews*, Vol. 5, 1986.

Manis mastodon site: Eric Bergland and Jerry Marr, *Prehistoric Life on the Olympic Peninsula*, Pacific Northwest National Parks and Forests Association, Seattle, 1988.

Paleobotany of the Olympics: Cathy Whitlock, "Vegetational and Climatic History of the Pacific Northwest During the Last 20,000 Years: Implications for Understanding Present-Day Biodiversity," *The Northwest Environmental Journal*, Vol. 8,1992.

Olympics as Ice Age refuge for plants: Nelsa Buckingham, "The Uniqueness of the Olympics," unpublished manuscript, Olympic National Park; also Nelsa Buckingham, E. Schreiner, T. Kaye, J. Burger, and E. Tisch, *Flora of the Olympic Peninsula, Washington*, Northwest Interpretive Association, Seattle, 1995.

Ice Age effects on Olympic fauna: Douglas B. Houston, B. Moorhead and E. Schreiner, *Mountain Goats in Olympic National Park: Biology and Management of an Introduced Species*, Scientific Monograph NPS/NROLYM/NRSM-94/25, U.S. Department of the Interior National Park Service, Denver, 1994.

Little Ice Age: Jan A. Henderson, D. Peter, R. Lesher, and D. Shaw, *Forested Plant Associations of the Olympic National Forest*, USDA Forest Service, Pacific Northwest Region, R6 Ecol. Technical Paper 001-88, Portland, Oregon, 1989.

History of Blue Glacier: Calvin Heusser, "Variations of Blue, Hoh, and White Glaciers during Recent Centuries," *Arctic*, Vol. 10, 1957; also Calvin Heusser, "Quaternary Vegetation, Climate, and Glaciation of the Hoh River Valley, Washington," *Geological Society of America Bulletin*, Vol. 85, 1974.

Recent behavior of Blue Glacier: Richard Spicer, "Recent Variations of Blue Glacier, Olympic Mountains, Washington, USA," *Arctic and Alpine Research*, Vol 21, No. 1, 1989.

Current mass wasting studies on Blue Glacier: Howard Conway, University of Washington, personal communication.

Chapter 3 The High-Country Year
An excellent and informative field guide to the Olympics' subalpine areas as well as its lowland forest communities is Daniel Matthews' *Cascade-Olympic Natural History, A Trailside Reference*, Raven Editions/Portland Audubon

Society, Portland, 1988. A fascinating comparison of Olympic's alpine areas with others in North America is Ann Zwinger and Beatrice Willard's *Land Above the Trees, A Guide to American Alpine Tundra*, Harper and Row, New York, 1972.

Ecology of subalpine forests: Stephen Arno and Ramona Hammerly, *Northwest Trees*, The Mountaineers, Seattle, 1977; Jan Henderson, D. Peter, R. Lesher, and D. Shaw, *Forested Plant Associations of the Olympic National Forest*, USDA Forest Service, Pacific Northwest Region, 1989; and R. W. Fonda and L. Bliss "Forest Vegetation of the Montane and Subalpine Zones, Olympic Mountains, Washington," *Ecological Monographs*, 39, 1969, and R. W. Fonda, "Ecology of Alpine Timberline in Olympic National Park," Proceedings, Conference on Scientific Research in National Parks, I, 1976.

Subalpine tree advancement into meadows: Andrea Woodward, University of Washington, personal communication.

Climate change as indicated by subalpine forests: Andrea Woodward, Michael Gracz, and Edward Schreiner, "Climatic Effects on Establishment of Subalpine Firs *(Abies lasiocarpa)* in Meadows of the Olympic Mountains," *Northwest Environmental Journal*, Vol. 7, No. 2, 1991.

Alpine and subalpine wildflowers: Charles Stewart, *Wildflowers of the Olympics and Cascades*, Nature Education Enterprises, Port Angeles, Washington, 1988.

Subalpine plant communities: Richard Kuramoto and L. Bliss, "Ecology of Subalpine Meadows in the Olympic Mountains," *Ecological Monographs*, 40, Summer 1970.

Alpine meadow communities: Zwinger and Willard, 1972, and L. Bliss, "Alpine Community Patterns in Relation to Environmental Parameters," from K. Greenridge, *Essays in Plant Geography and Ecology*, Nova Scotia Museum, Halifax, 1969.

Small mammals of alpine and subalpine areas: James Reichel, "Ecology of Pacific Northwest Alpine Mammals," doctoral thesis, Washington State University, 1984, and Earl Larrison and Amy Fisher, *Mammals of the Northwest*, Seattle Audubon Society, 1966.

Ecology and behavior of the Olympic marmot: David Barash, "The Behavior of the Olympic Marmot," master's thesis, University of Wisconsin, Madison, 1967; David Barash, "The Social Biology of the Olympic Marmot," doctoral thesis, University of Wisconsin, Madison, 1970; also Roger del Moral, "The Impact of the Olympic Marmot on Subalpine Vegetation Structure," *American Journal of Botany*, 71(9), 1984; also William Wood, "Habitat Selection and Energetics of the Olympic Marmot," master's thesis, Western Washington State College, 1973.

Mammals common in the Cascades but historically missing from the Olympics: Victor B. Scheffer, *Mammals of the Olympic Peninsula*, Society for Northwestern Vertebrate Biology, Olympia, Washington, 1995; and W. W. Dalquest, *Mammals of Washington*, University of Kansas Press, Lawrence, 1948.

Mountain goats: *Goats in Olympic National Park: Environmental Impact Statement for Mountain Goat Management*, USDI National Park Service, Port Angeles, Washington, 1995.

Chapter 4 The Rain Forest

An extensive and up-to-date treatment of the ecology of the Olympic rain forest is Ruth Kirk and Jerry Franklin's *The Olympic Rain Forest, An Ecological Web*, University of Washington Press, Seattle, 1992. A handy field guide to rain forest trees and shrubs is C. P. Lyons and Bill Merilees's *Trees, Shrubs and Flowers to Know in Washington and British Columbia*, Lone Pine Publishing, Vancouver, 1995.

Factors leading to the rain forest's development: Kirk and Franklin, 1992.

Sitka spruce–western hemlock forest type: Jerry F. Franklin and C. T. Dyrness, *Natural Vegetation of Oregon and Washington*, USDA Forest Service Pacific Northwest Range and Experiment Station, General Technical Report PNW-8, 1973.

Natural history of rain forest trees: Steven Arno and R. Hammerly, *Northwest Trees*, The Mountaineers, Seattle, 1977; also Daniel Matthews, *Cascade-Olympic Natural History, A Trailside Reference*, Raven Editions/Portland Audubon Society, 1988.

River terrace communities: R. W. Fonda, "Forest Succession in Relation to River Terrace Development in Olympic National Park, Washington," *Ecology*, 55:5, 1974.

Ecological function of downed logs in forests: Chris Maser, R. F. Tarrant, J. M. Trappe, and J. F. Franklin, *From the Forest to the Sea: A Story of Fallen Trees*, Pacific Northwest Research Station, USDA Forest Service, Portland, Oregon, 1988.

Nurse logs: Kirk and Franklin, 1992.

Canopy ecology: Bruce McCune, "Eradiants in Epiphyte Biomass in Three Pseudotsuga-Tsuga Forests of Different Ages in Oregon and Washington," *The Brycologist*, 96(3), 1993.

Rain forest trees rooting into epiphyte mats: Nalini M. Nadkarni, "Roots that Go Out on a Limb," *Natural History*, Vol. 94, 1985.

Elk in the Olympics: Bruce B. Moorhead, *The Forest Elk, Roosevelt Elk in Olympic National Park*, Northwest Interpretive Association, Seattle, 1994; also D. B. Houston, E. G. Schreiner, B. B. Moorhead, "Elk in Olympic National Park: Will They Persist Over Time?" *Natural Areas Journal*, Volume 10(1), 1990.

Elk–rain forest interactions: K. Jenkins and E. Starkey, "Habitat Use by Roosevelt Elk in Unmanaged Forests of the Hoh Valley, Washington," *Journal of Wildlife Management*, 48, 1984; also A. Woodward, E. G. Schreiner, D. B. Houston, and B. B. Moorhead, "Ungulate-Forest Relationships in Olympic National Park: Retrospective Exclosure Studies," *Northwest Science*, Vol. 68, no. 2, 1994.

History of elk management in the Olympics: Jenkins and Starkey, 1984.

Elk in the east-side valleys: Gregory L. Schroer, "Seasonal Movements and Distribution of Migratory Roosevelt Elk in the Olympic Mountains," master's thesis, Oregon State University, Corvallis, 1986.

Cougars: Maurice G. Hornocker, "Learning to Live with Mountain Lions," *National Geographic*, July 1992.

Other mammals: Victor B. Scheffer, *Mammals of the Olympic Peninsula*, Society for Northwestern Vertebrate Biology, Olympia, Washington, 1995.

Wolves in the Olympics: Peter Dratch, B. Johnson, L. Leigh, D. Levkoy, D. Milne, R. Read, R. Selkirk, and C. Swanberg, "A Case Study for Species Reintroduction: The Wolf in Olympic National Park" (unpublished report), The Evergreen State College, Olympia, Washington, 1975; also L. David Mech, "The Challenge and Opportunity of Recovering Wolf Populations," *Conservation Biology*, Vol. 9, No. 2, April, 1995.

Chapter 5 The Old-Growth Forest Community

The most comprehensive book for the general reader on the ecology of old-growth forests in the Northwest is Elliot Norse's *Ancient Forests of the Pacific Northwest*, The Wilderness Society and Island Press, Washington, D.C., 1990.

Spotted owls and old-growth forests: Norse, 1990.

Wildlife species of the peninsula: Fred Sharpe, "Olympic Wildlife Checklist," Northwest Interpretive Association, Seattle, 1992.

"Scuzz": Kirk and Franklin, 1992.

Spotted owls and flying squirrels: Norse, 1990.

Rodents and fungi in the forest ecosystem: Chris Maser, *Forest Primeval, The Natural History of An Ancient Forest*, Sierra Club Books, San Francisco, 1989.

Structure and ecological function of old-growth forests: Jerry F. Franklin, et al., *Ecological Characteristics of Old-Growth Douglas-Fir Forests*, USDA Forest Service Pacific Northwest Range and Experiment Station, General Technical Report PNW-118, 1981.

Wildlife use of old-growth forests: Norse, 1990; also Larry D. Harris, *The Fragmented Forest, Island Biogeography Theory and the Preservation of Biotic Diversity*, University of Chicago Press, Chicago, 1984; also Bruce B. Moorhead, "A Summary of the U.S. Forest Service Symposium of Wildlife in Old-growth Douglas-fir Forests of the Pacific Northwest, March, 1989, Portland, Oregon," unpublished report, Olympic National Park.

Marbled murrelets: Sharon Levy, "A Closer Look: Marbled Murrelet," *Birding*, December 1993; also Susan McCarthy, "A Seabird's Secret Life is Revealed—50 Miles Inland," *Smithsonian*, Vol. 2, No. 9, 1993.

Birds of the old-growth forest: Norse, 1990; also Fred Sharpe, *Olympic Peninsula Birds: The Songbirds*, unpublished manuscript.

Snags and other old-growth habitats: E. Reade Brown, *Management of Wildlife and Fish Habitat in Forests of Western Washington and Oregon*, USDA Forest Service, 1985; also Norse, 1990.

Martins and fishers: Larrison and Fisher, 1976; Keith B. Aubrey and D. B. Houston, "Distribution and Status of the Fisher *(Martes pennanti)* in Washington," *Northwestern Naturalist*, Winter 1992; also D. B. Houston and D. E. Seaman, "Fisher in Olympic National Park," Research Note no. 2, Olympic National Park, 1983.

Forest vegetation types: Jan A. Henderson, D. Peter, R. Lesher, and D. Shaw, *Forested Plant Associations of the Olympic National Forest*, USDA Forest Service, Pacific Northwest Region, R6 Ecol. Technical Paper 001-88, Portland, Oregon, 1989; also Jerry F. Franklin and C. T. Dyrness, *Natural Vegetation of Oregon and Washington*, USDA Forest Service Pacific Northwest Range

and Experiment Station, General Technical Report PNW-8, 1973.
Plants of the lowland forests: Arno and Hammerly, 1977; also Matthews, 1988.
Montaine forests: R. W. Fonda and L. C. Bliss, "Forest Vegetation of the Montane and Subalpine Zones, Olympic Mountains, Washington," *Ecological Monographs* 39, Summer, 1969.
Rainshadow vegetation: Jerry Gorsline, "The Cultural Transformation of the Sequim Prairie," in Gorsline, *Shadows of Our Ancestors*, Empty Bowl, Port Townsend, 1992.
Fire history of Olympic forests: Henderson, et al., 1989.
Current old-growth on the Olympic Peninsula: Peter Morrison, *Ancient Forests of the Olympic National Forest, Analysis from a Historical and Landscape Perspective*, The Wilderness Society, Washington, D.C., 1989.

Chapter 6 The Lives of Olympic Rivers
A landmark book explaining the plight of wild salmon on the Olympic Peninsula is Bruce Brown's *Mountain in the Clouds, A Search for the Wild Salmon*, Simon and Schuster, New York, 1982. The natural history and biology of salmon is drawn from C. Croot and L. Margolis's *Pacific Salmon Life Histories*, University of British Columbia Press, Vancouver, 1991; Richard S. Wydoski and R. Whitney's *Inland Fishes of Washington*, University of Washington Press, Seattle, 1979; and Robert Steelquist's *Adopt-a-Stream Foundation Field Guide to the Pacific Salmon*, Sasquatch Books, Seattle, 1992.
Salmon stocks in Olympic National Park: D. B. Houston, "Anadromous Fish in Olympic National Park: A Status Report," Olympic National Park, 1983; also *1992 Washington State Salmon and Steelhead Stock Inventory*, Washington Department of Fisheries, Washington Department of Wildlife, and Western Washington Treaty Tribes, Olympia, 1992.
Salmon stocks at risk in peninsula rivers: Willa Nehlsen, J. Williams, and J. Lichatowich, "Pacific Salmon at the Crossroads: Stocks at Risk from California, Oregon, Idaho, and Washington," *Fisheries*, Vol. 16, No. 2, 1991; also Jim Lichatowich, "The Status of Anadromous Fish Stocks in the Streams of Eastern Jefferson County, Washington," Jamestown S'Klallam Tribe, Sequim, Washington, 1993.
Spawned salmon as important seasonal food for wildlife: C. J. Cederholm, D. Houston, and W. Scarlett, "Fate of Coho Salmon *(Oncorhynchus kisutch)* Carcasses in Spawning Streams," *Canadian Journal of Fisheries and Aquatic Sciences*, Vol. 46, 1989.
Salmon cycling nutrients to streams and forests: Robert E. Bilby, B. Fransen, and P. Bisson, "Incorporation of Nitrogen and Carbon from Spawning Coho Salmon into the Trophic System of Small Streams: Evidence from Stable Isotopes," *Canadian Journal of Fishereis and Aquatic Sciences* (in press).
Ecology of riparian areas: Art Oakley, et al., "Riparian Zones and Freshwater Wetlands," in E. Reade Brown, *Management of Wildlife and Fish Habitat in Forests of Western Washington and Oregon*, USDA Forest Service, 1985; also Tim McNulty, *Washington's Wild Rivers, The Unfinished Work*, The Mountaineers, Seattle, 1990.
Downed logs provide structure for streams: Chris Maser, R. Tarrant, J. Trappe,

and J. Franklin, *From the Forest to the Sea: A Story of Fallen Trees*, USDA Forest Service General Technical Report PNW-GTR-229, Portland, Oregon, 1988.

Riparian-associated wildlife: Maser, et al., 1988.

Harlequin ducks: Douglas H. Chadwick, "Bird of White Waters," *National Geographic*, November, 1993; also "Cavity Nesting Harlequin Ducks in the Pacific Northwest," *Wilson Bulletin*, Vol. 105, no. 4, 1993.

Dippers, kingfishers, and other river-associated birds: Paul R. Ehrlich, D. Dobkin, and D. Wheye. *The Birder's Handbook*, Simon and Schuster, New York, 1988; also E. A. Kitchin, *Birds of the Olympic Peninsula*, Olympic Stationers, Port Angeles, 1949; Roger Tory Peterson, *A Field Guide to Western Birds*, Houghton Mifflin Company, Boston, 1990; also Fred A. Sharpe, *Olympic Peninsula Birds, The Songbirds*, unpublished manuscript, Olympic National Park.

Amphibians of the Olympics: William P. Leonard, H. Brown, L. Jones, K. McAllister, and R. Storm, *Amphibians of Washington and Oregon*, The Seattle Audubon Society, 1993; also Bruce B. Bury, "Amphibians and Reptiles of the Olympic National Park, Washington," Preliminary Report, 1982; also Bruce B. Moorhead, "Report on Northwest Amphibians Symposium, Astoria, Oregon, March, 1993," Olympic National Park science files.

Worldwide amphibian declines: David B. Wake, "Declining Amphibian Populations," *Science*, Vol. 253, 1991.

Restoration of the Elwha River: Cat H. Hoffman, "The Elwha Issue: A Fish Problem that Just Won't Die," *The George Wright Forum*, Volume 9, No. 2, 1992; also *The Elwha Report*, U.S. Department of the Interior, Department of Commerce, and Elwha S'Klallam Tribe, 1994; *Elwha River Ecosystem Restoration, Final Environmental Impact Statement*, U.S. Department of the Interior, National Park Service, 1995; also Robert Wunderlich, Brian Winter, and John Meyer, "Restoration of the Elwha River Ecosystem and Anadromous Fisheries," *Bethesda* 19 (8), 1994.

Chapter 7 Life at the Edge of Land and Sea

The chapter on the ecology of Olympic's intertidal zones was assembled from the work of Megan Dethier of the University of Washigton, along with a number of published sources. Gloria Snively's *Exploring the Seashore in British Columbia, Washington and Oregon*, Gordon Soules Book Publishers Ltd. and Pacific Search Press, Vancouver and Seattle, 1978, is probably the best introduction to intertidal zones and organisms; she uses illustrations and common as well as scientific names for each species and arranges subject matter by life zones. Eugene N. Kozloff's *Seashore Life of the Northern Pacific Coast*, University of Washington Press, Seattle, 1983, presents a more detailed look at intertidal organisms, and Edward Ricketts, J. Calvin and J. Hedgpeth's *Between Pacific Tides* (fourth edition), Stanford University Press, Stanford, 1968, is the classic in the field. Anne Wertheim's *The Intertidal Wilderness*, Sierra Club Books, San Francisco, 1984, offers exquisite underwater photography as well as keen ecological insight. Muriel L. Guberlet's *Seaweeds at Ebb Tide*, University of Washington Press, Seattle, 1956, and *Animals of the Seashore* (revised edition), Binfords and Mort, Portland, 1962, though somewhat dated, are both excellent introductions to marine organisms. Ruth Kirk's *The Olympic Seashore*, Pacific Northwest National Parks Association, Seattle, 1962, and David Hooper's *Exploring Washington's Wild Olympic Coast*, The Mountaineers, Seattle, 1993, are popular hiking guides to the coast that contain some useful natural history as well.

Factors leading to the diversity of coastal environments: Megan N. Dethier, "A Survey of Intertidal Communities of the Pacific Coast Area of Olympic National Park, Washington," Olympic National Park unpublished report, 1988; also Wertheim, 1984.

Coastal geology: Weldon W. Rau, *Geology of the Washington Coast Between the Hoh and Quillayute Rivers*, Washington Department of Natural Resources Bulletin No. 72, Olympia, 1980; and Tabor, 1986.

Shape of the Olympic coast: Thomas Terich and M. L. Schwartz, "A Geomorphic Classification of Washington State's Pacific Coast, Shore and Beach," 1981.

Origin of Point of Arches rocks: Tim McNulty and P. O'Hara, *Olympic National Park, Where the Mountain Meets the Sea*, Woodlands Press, Del Mar, 1984.

Currents and tides: Thomas Carefoot, *Pacific Seashores*, University of Washington Press, Seattle, 1977.

Species succession in the intertidal: R. T. Paine, and S. A. Levin, "Intertidal Landscapes: Disturbance and the Dynamics of Pattern," *Ecological Monographs*, Vol. 51, No. 2, 1981.

Mussel bed communities and sea star–mussel predation: R. T. Paine, "Intertidal Community Structure: Experimental Studies on the Relationship between a Dominant Competitor and Its Principal Predator," *Oecologia*, Vol. 15, 1974; also Wertheim, 1984.

Interactions among sea anemones: Kenneth Sebens, "Population Dynamics and Habitat Suitability of the Intertidal Sea Anemones *Anthopleura elegantissima* and *A. xanthogrammica*," *Ecological Monographs*, Vol. 53, 1983.

Herbivores and seaweed community structure: Dethier, 1988.

Chapter 8 Seabirds and Mammals of the Outer Coast

The natural histories of seabirds and mammals on the coast was drawn from a number of published sources as well as interviews with several biologists conducting surveys and research in this area. Among books available on Washington's avifauna, Stanley Jewett, W. Taylor, W. Shaw, and J. Aldrich's *Birds of Washington State*, University of Washington Press, Seattle, 1953, was essential reading for its site-specific detail and historical perspective. E. A. Kitchin, *Birds of the Olympic Peninsula*, Olympic Stationers, Port Angeles, Washington, 1949, was also extremely useful. Paul Ehrlich, D. Dobkin, and D. Wheye's *The Birder's Handbook*, Simon and Schuster, New York, 1988, filled in biological and natural history details as well as conservation status for several species, as did C. J. Guiguet's, *The Birds of British Columbia, (9) Diving Birds and Tube-nosed Swimmers*, British Columbia Provincial Museum, Victoria, 1971. I relied primarily on Roger Tory Peterson's *A Field Guide to Western Birds*, Houghton Mifflin Company, Boston, 1990, for identification and descriptions.

For marine mammals, Tony Angell and K. Balcomb's *Marine Birds and Mammals of Puget Sound*, Puget Sound Books, Seattle, 1982, is an excellent introduction to Northwest marine life. Richard Osborne, J. Calambokidis, and E. Dorsey's *A Guide to Marine Mammals of Greater Puget Sound*, Island Publishers, Anacortes, Washington, 1988, is a clear and insightful introduction to the area's marine mammals, and Delphine Haley's compilation, *Marine Mammals of Eastern North Pacific and Arctic Waters*, Pacific Search Press, Seattle, 1986, contains excellent articles on the region's sea mammals by leading researchers. A pioneering paper on sea mammals in Washington is Victor Scheffer and J. Slipp's monograph, "The Whales and Dolphins of Washington State," in *The American Midland Naturalist*, Vol. 39, 1948. A good overview of Washington's coastal areas is Robert Steelquist's *Washington's Coast*, American Geographic Publishing, Helena, Montana, 1987.

History of shipwrecks on the Olympic coast: James Gibbs, *Shipwrecks of the Pacific Coast*, Binfords and Mort, Portland, 1962.

William Dawson's early work on coastal seabird colonies: Steelquist, 1987.

"Quileute": There are different spellings for the Native American tribe (Quileute) and the river (Quillayute).

Status of bald eagles: Kelly McAllister, T. Owens, L. Leschner, and Eric Cummins, "Distribution and Productivity of Nesting Bald Eagles in Washington, 1981–1985," *The Murrelet*, 67, 1986; and Anita McMillan, Washington Department of Fish and Wildlife, personal communication.

Current status of Pacific seabirds: Kees Vermeer, K. Briggs, K. Morgan, and D. Siegel-Causey, *The Status, Ecology, and Conservation of Marine Birds of the North Pacific*, Canadian Wildlife Service, Ottawa, 1993, also David G. Ainley, W. Sydeman, S. Hatch, and U. Wilson, "Seabird Population Trends Along the West Coast of North America: Causes and the Extent of Regional Concordance," *Studies in Avian Biology*, No. 15, 1994; and Ulrich Wilson, U.S. Fish and Wildlife Service, personal communication.

Impacts of El Niño on seabird colonies: Ulrich Wilson, "Responses of Three Seabird Species to El Niño Events and Other Warm Episodes on the Washington Coast, 1979–1990," *Condor*, 1991.

Natural history and recovery of gray whales: Mary Lou Jones and S. Swartz, *The Gray Whale, Eschrichtius robustus*, Academic Press, Orlando, Florida, 1984; also Ben Bennet, *The Oceanic Society Field Guide to the Gray Whale*, Sasquatch Books, Seattle, 1989.

Overview of current research on marine mammals: Bruce Moorhead, "Notes on Marine Mammals from British Columbia/Washington Wildlife Society Conference, 1993," unpublished notes, Olympic National Park.

Status of harbor seals: Harriet Huber, S. Jeffries, R. Brown, and R. DeLong, "Abundance of Harbor Seals in Washington and Oregon," National Marine Mammal Laboratory, Seattle, 1992.

Overview of sea otters: Roy Nickerson, *Sea Otters, A Natural History Guide*, Chronicle Books, San Francisco, 1989.

Sea otter recovery on the Olympic coast: C. Edward Bowlby, B. Troutman, and S. Jeffries, "Sea Otters in Washington: Distribution, Abundance, and Activity Patterns," Washington Department of Wildlife, Olympia, 1988; also R. Jameson, "Olympic Coast Sea Otter Study: 1994 Progress Report," USDI National Biological Service, Corvallis, Oregon.

Reasons for limited sea mammal breeding on Olympic coast: Steve Jeffries, Washington Department of Fish and Wildlife, personal communication.

Effects of oil on seabirds: R. Glenn Ford, D. Varoujean, D. Warrick, W. Williams, D. Lewis, C. Hewett, and J. Casey, "Seabird Mortality Resulting from the Nestucca Oil Spill Incident," Washington Department of Wildlife Report, Olympia, 1991.

Impacts of oil on intertidal species: Megan Dethier, "The Effects of an Oil Spill and Freeze Event on Intertidal Community Structure in Washington," U.S. Department of the Interior, Minerals Management Service Report, Camarillo, California, 1991.

Chapter 9 Footprints on the Land

An excellent overview of early archeological findings on the peninsula is Eric Bergland and J. Marr's *Prehistoric Life on the Olympic Peninsula*, Pacific Northwest National Parks and Forests Association, Seattle, 1988. Ethnographic information on peninsula tribal peoples is drawn largely from the Smithsonian Institution's excellent series, *Handbook of North American Indians; Volume 7, Northwest Coast*, edited by Wayne Suttles, Smithsonian Institution, Washington, 1990.

Manis mastodon discovery: Carl Gustafson, D. Gilbrow, and R. Daugherty, "The Manis Mastodon Site: Early Man on the Olympic Peninsula," *Canadian Journal of Archaeology*, No. 3, 1979.

Investigation of the Manis site hunting camp: Delbert Gilbrow, personal communication.

Prehistoric migrations to the peninsula: Eric Bergland, "Summary Prehistory and Ethnography of Olympic National Park, Washington," National Park Service, Pacific Northwest Region, Seattle, 1983.

Slab Camp site: James Gallison, "Slab Camp: An Early to Middle Holocene Olcott Complex in the Eastern Olympic Mountains of Washington," doctoral thesis, Washington State University Department of Anthroloplogy, 1994.

Coastal Olcott pattern: Ruth Kirk and R. Daugherty, *Exploring Washington Archaeology*, University of Washington Press, Seattle, 1978, and David Conca, Olympic National Park, personal communication.

Deer Park site: Gary C. Wessen, "Archaeological Survey and Testing Activities at and near Deer Park (45CA257), Olympic National Park, Washington," unpublished report, USDI National Park Service, Pacific Northwest Region, 1992.

Seven Lakes Basin site: David Conca, "Soleduck Watershed High Country Archaeology Synopsis," Olympic National Park, 1995.

Glenrose Cannery and Hoko River sites: Kirk, 1978, and Bergland, 1988.

Other early maritime sites: Gary Wessen, "Prehistory of the Ocean Coast of Washington," in Wayne Suttles, *Handbook of North American Indians; Volume 7, Northwest Coast*, Smithsonian Institution, Washington, 1990.

Discovery and early excavation of Ozette site: Ruth Kirk, *Hunters of the Whale*, Morrow, New York, 1974.

Ozette archaeological findings: Kirk, 1978, and Stephan Samuels, *Ozette*

Archaeological Project Research Reports, Vol. 1, U.S. Department of Anthropology Reports of Investigations 63, National Park Service, Pacific Northwest Regional Office, Seattle, 1991; also Paul Gleeson, Olympic National Park, personal communication.

Pre-contact Makah village life: Mary Parker Pascua, "Ozette, A Makah Village in 1491," *National Geographic*, October, 1991.

Makah ethnography: Ann M. Renker and E. Gunther, "Makah," in Suttles, 1990; also James Swan, *The Indians of Cape Flattery*, Smithsonian Institution, Washington, facsimile reproduction, the Shorey Book Store, Seattle, 1974.

Northwest Coast cultural adaptations: Wayne Suttles, "Introduction to the Northwest Coast," in Suttles, 1990; also Philip Drucker, *Indians of the Northwest Coast*, McGraw Hill Book Co., New York, 1955.

The potlatch: Ruth Kirk, *Tradition and Change on the Northwest Coast*, University of Washington Press, Seattle, 1986, and Suttles, 1990.

Cedar technology: Hillary Stewart, *Cedar, Tree of Life to the Northwest Coast Indians*, Douglas and McIntyre, Vancouver, 1984.

Traditional use of native plants: Erna Gunther, *Ethnobotany of Western Washington*, University of Washington Press, Seattle, 1973.

Quileute ethnography: James V. Powell, "Quileute," in Suttles, 1990.

Quinault and Queets ethnography: Yvonne Hajda, "Southwestern Coast Salish," in Suttles, 1990.

S'Klallam ethnography: Wayne Suttles, "Central Coast Salish," in Suttles, 1990.

Twana/Skokomish ethnography: Wayne Suttles and B. Lane, "Southern Coast Salish," in Suttles, 1990.

Treaties and fishing rights: Fay G. Cohen, *Treaties on Trial*, University of Washington Press, Seattle, 1986.

Chapter 10 Protection for Olympic's Wildlands

The story of the long and contentious struggle to create Olympic National Park, as well as the ongoing effort to protect the park's valuable old-growth forests, is told in Carsten Lien's excellent study, *Olympic Battleground, The Power Politics of Timber Preservation*, Sierra Club Books, San Francisco, 1991. A useful anthology that excerpts a number of primary texts on the park's history is Nancy Beres, M. Chandler, and R. Dalton's *Island of Rivers*, Pacific Northwest National Parks and Forests Association, Seattle, 1988. A wonderful book on the early exploration of the Olympics is Robert Wood's *The Land that Slept Late, The Olympic Mountains in Legend and History*, The Mountaineers, Seattle, 1995.

Wickersham and O'Neil on the Skokomish River: Robert L. Wood, *Men, Mules and Mountains, Lieutenant O'Neil's Olympic Expeditions*, The Mountaineers, Seattle, 1976.

Account of the Press expedition: Robert L. Wood, *Across the Olympic Mountains, The Press Expedition, 1889-90*, University of Washington Press, Seattle, 1967.

Early expeditions into the Olympics: Gail Evans, *Historic Resource Study, Olympic National Park*, Cultural Resources Division, Pacific Northwest

Region, National Park Service, Seattle, Washington, 1983.

Wickersham's park proposal: James Wickersham, "A National Park in the Olympics . . . 1890," *The Living Wilderness,* Summer/Fall, 1961.

Early conservation efforts in the U.S.: Roderick Nash, *Wilderness and the American Mind,* Yale University Press, New Haven, 1969.

Creation and reduction of the Olympic Forest Reserve: Lien, 1991.

Local efforts to reduce the Reserve: Thomas T. Aldwell, *Conquering the Last Frontier,* Superior Publishing Company, Seattle, 1950.

Chris Morgenroth and the homesteaders' perspective on Forest Reserve reductions: Chris Morgenroth, *Footprints in the Olympics, an Autobiography,* Ye Galleon Press, Fairfield, Washington, 1991.

Gifford Pinchot and the formative years of the Forest Service: Gifford Pinchot, *Breaking New Ground,* Harcourt Brace Jovanovich, Inc., New York, 1945.

Campaign for a national park: Lien, 1991.

Power struggle between the park service and forest service: Ben W. Twight, *Organizational Values and Political Power: The Forest Service Versus the Olympic National Park,* Pennsylvania State University Press, University Park, 1983.

Army Spruce Division and World War I: Tim McNulty, "The Spruce Railroad Trail, Olympic National Park," *Peninsula Magazine,* Winter, 1989.

Emergency Conservation Committee and the national campaign for a park: Lien, 1990.

"The peninsula affords the last opportunity . . . ": Willard Van Name, "The Proposed Olympic National Park," Emergency Conservation Committee pamphlet, 1934.

FDR and the creation of the park: Lien, 1991.

Salvage logging in Olympic National Park: Lien, 1991.

Citizen activism following creation of the Park: Polly Dyer, "40 Years of Olympic Park Associates," *Voice of the Wild Olympics,* October, 1988.

Definition of ecosystem management: T. Clark and D. Zaunbrecher, "The Greater Yellowstone Ecosystem, The Ecosystem Concept in Natural Resource Policy and Management," *Renewable Resources Journal,* 11, 1987.

Perspectives on ecosystem management: R. Edward Grumbine, *Ghost Bears, Exploring the Biodiversity Crisis,* Island Press, Washington, D.C., 1992; also James Agee and D. Johnson, *Ecosystem Management for Parks and Wilderness,* Pacific Northwest Region, National Park Service, Seattle, 1988.

Components of the Olympic ecosystem: Michael Smithson and P. O'Hara, *Olympic, Ecosystems of the Peninsula,* America and World Geographic Publishing, Helena, Montana, 1993.

Strategies for ecosystem management in the Olympics: Tim McNulty, "The Olympic Ecosystem," *Voice of the Wild Olympics,* October, 1988.

UNESCO's Biosphere Reserve and World Heritage programs: Michel Batisse, "Development and Implementation of the Biosphere Reserve Concept and Its Applicability to Coastal Regions," *Environmental Conservation,* Vol. 17, No. 2, 1990.

UNESCO quote: "Comments of Dr. Michel Batisse, June 29, 1982," in Beres, et al., 1988.

BIBLIOGRAPHY

Agee, James, and D. Johnson. 1988. *Ecosystem Management for Parks and Wilderness*. Seattle: Pacific Northwest Region, National Park Service.

Aldwell, Thomas T. 1950. *Conquering the Last Frontier*. Seattle: Superior Publishing Company.

Angell, Tony, and Kenneth C. Balcomb. 1982. *Marine Birds and Mammals of Puget Sound*. Seattle: Washington Sea Grant/Puget Sound Books.

Arno, Stephen, and Ramona Hammerly. 1977. *Northwest Trees*. Seattle: The Mountaineers.

Barash, David. 1989. *Marmots: Social Behavior and Ecology*. Stanford: Stanford University Press.

Bennet, Ben. 1989. *The Oceanic Society Field Guide to the Gray Whale*. Seattle: Sasquatch Books.

Beres, Nancy, Mitzi Chandler, and Russ Dalton (eds.). 1988. *Island of Rivers: An Anthology Celebrating 50 Years of Olympic National Park*. Seattle: Pacific Northwest National Parks and Forests Association.

Bergland, Eric O., and Jerry Marr. 1988. *Prehistoric Life on the Olympic Peninsula*. Seattle: Pacific Northwest National Parks and Forests Association.

Brown, Bruce. 1982. *Mountain in the Clouds: A Search for the Wild Salmon*. New York: Simon and Schuster.

Brown, E. Reade (ed.). 1985. *Management of Wildlife and Fish Habitat in Forests of Western Washington and Oregon*. Washington, D.C.: USDA Forest Service.

Buckingham, Nelsa M., Edward Schreiner, Thomas Kaye, Janis Burger, and Edward Tisch. 1995. *Flora of the Olympic Peninsula, Washington*. Seattle: Northwest Interpretive Association.

Burns, Robert. 1985. *The Shape of Puget Sound*. Seattle: Washington Sea Grant/Puget Sound Books.

Burt, William, and R. Grossenheider. 1976. *A Field Guide to the Mammals*. Boston: Houghton Mifflin Company.

Carefoot, Thomas. 1977. *Pacific Seashores*. Seattle: University of Washington Press.

Cederholm, C. J., D. Houston, and W. Scarlett. 1989. "Fate of Coho Salmon *(Oncorhynchus kisutch)* Carcasses in Spawning Streams." *Canadian Journal of Fisheries and Aquatic Sciences*, Vol. 46.

Chadwick, Douglas H. 1993. "Bird of White Waters." *National Geographic*, November.

Cohen, Fay G. 1986. *Treaties on Trial*. Seattle: University of Washington Press.

Cowan, Ian M., and C. J. Guiguet. 1965. *The Mammals of British Columbia*. Victoria: British Columbia Provincial Museum.

Croot, C., and L. Margolis. 1991. *Pacific Salmon Life Histories*. Vancouver: University of British Columbia Press.

Dalquest, W. W. 1948. *Mammals of Washington*. Lawrence: University of Kansas Press.

del Moral, Roger. 1984. "The Impact of the Olympic Marmot on Subalpine Vegetation Structure." *American Journal of Botany*, 71(9), 1228–1236.

Dethier, Megan N. 1988. "A Survey of Intertidal Communities of the Pacific

Coast Area of Olympic National Park, Washington." Olympic National Park unpublished report.

Dietrich, William. 1992. *The Final Forest: The Battle for the Last Great Trees of the Pacific Northwest*. New York: Simon and Schuster.

Dodwell, Arthur, and Theodore F. Rixon. 1902. *Forest Conditions in the Olympic Forest Reserve, Washington*. U.S. Geological Survey Professional Paper No. 7. Washington D.C.: U.S. Government Printing Office.

Drucker, Philip. 1955. *Indians of the Northwest Coast*. New York: McGraw Hill Book Co.

Ehrlich, Paul, D. Dobkin, and D. Wheye. 1988. *The Birder's Handbook: A Field Guide to the Natural History of North American Birds*. New York: Simon and Schuster.

Evans, Gail. 1983. *Historic Resource Study, Olympic National Park*. Seattle: National Park Service, Pacific Northwest Region.

Fonda, R. W. 1974. "Forest Succession in Relation to River Terrace Development in Olympic National Park, Washington." *Ecology*, 55:5.

Fonda, R. W., and L. Bliss. 1969. "Forest Vegetation of the Montane and Subalpine Zones, Olympic Mountains, Washington." *Ecological Monographs*, 39.

Franklin, Jerry F., and Dyrness, C. T. 1973. *Natural Vegetation of Oregon and Washington*. Portland: USDA Forest Service Pacific Northwest Range and Experiment Station, General Technical Report PNW-8.

Franklin, Jerry F. et al. 1981. *Ecological Characteristics of Old-Growth Douglas-Fir Forests*. Portland: USDA Forest Service Pacific Northwest Range and Experiment Station, General Technical Report PNW-118.

Gibbs, James. 1962. *Shipwrecks of the Pacific Coast*. Portland: Binfords and Mort.

Gilkey, Helen M., and L. J. Dennis. 1975. *Handbook of Northwestern Plants*. Corvallis: Oregon State University.

Gorsline, Jerry. 1992. *Rainshadow; Archibald Menzies and the Botanical Exploration of the Olympic Peninsula*. Port Townsend, Washington: Jefferson County Historical Society.

Gorsline, Jerry (ed.) 1992. *Shadows of Our Ancestors: Readings in the History of Klallam-White Relations*. Port Townsend, Washington: Empty Bowl.

Grumbine, R. Edward. 1992. *Ghost Bears: Exploring the Biodiversity Crisis*. Washington, D.C.: Island Press.

Guberlet, Muriel L. 1956. *Seaweeds at Ebb Tide*. Seattle: University of Washington Press.

———. 1962. *Animals of the Seashore* (rev. ed.). Portland: Binfords and Mort.

Guiguet, C. J. 1971 *The Birds of British Columbia: Diving Birds and Tube-nosed Swimmers*. Victoria: British Columbia Provincial Museum.

Gunther, Erna. 1973. *Ethnobotany of Western Washington*. Seattle: University of Washington Press.

Haley, Delphine (ed). 1986. *Marine Mammals of Eastern North Pacific and Arctic Waters*. Seattle: Pacific Search Press.

Harris, Larry D. 1994. *The Fragmented Forest: Island Biogeography Theory and the Preservation of Biotic Diversity*. Chicago: University of Chicago Press.

Harthill, Marion P., and I. O'Connor. 1975. *Common Mosses of the Pacific Northwest*. Healdsburg, California: Naturegraph Publishers.

Henderson, Jan A., David Peter, Robin Lesher, and David Shaw. 1989. *Forested Plant Associations of the Olympic National Forest*. Portland: USDA Forest Service, Pacific Northwest Region, R6 Ecol. Technical Paper 001-88.

Hitchcock, C. Leo, and A. Cronquist. 1973. *Flora of the Pacific Northwest*. Seattle: University of Washington Press.

Hooper, David. 1993. *Exploring Washington's Wild Olympic Coast*. Seattle: The Mountaineers.

Hornocker, Maurice G. 1992. "Learning to Live with Mountain Lions." *National Geographic*, July.

Houston, D. B., E. G. Schreiner, B. B. Moorhead. 1990. "Elk in Olympic National Park: Will They Persist Over Time?" *Natural Areas Journal*, Vol. 10(1).

Houston, Douglas B., B. Moorhead, and E. Schreiner. 1994. *Mountain Goats in Olympic National Park: Biology and Management of an Introduced Species*. Denver: Scientific Monograph NPS/NROLYM/NRSM-94/25, USDI National Park Service.

Hult, Ruby El. 1954. *The Untamed Olympics: The Story of the Olympic Peninsula*. Portland: Bindfords and Mort.

Jenkins, K., and E. Starkey. 1984. "Habitat Use by Roosevelt Elk in Unmanaged Forests of the Hoh Valley, Washington." *Journal of Wildlife Management*, 48.

Jewett, Stanley, W. Taylor, W. Shaw, and J. Aldrich. 1953. *Birds of Washington State*. Seattle: University of Washington Press.

Jones, Mary Lou, and S. Swartz. 1984. *The Gray Whale, Eschrichtius robustus*. Orlando, Florida: Academic Press.

Kirk, Ruth. 1962. *The Olympic Seashore*. Seattle: Pacific Northwest National Parks Association.

———. 1974. *Hunters of the Whale*. New York: Morrow.

———. 1986. *Tradition and Change on the Northwest Coast*. Seattle: University of Washington Press.

Kirk, Ruth, and J. Franklin. 1992. *The Olympic Rain Forest: An Ecological Web*. Seattle: University of Washington Press.

Kirk, Ruth, and R. Daugherty. 1978. *Exploring Washington Archaeology*. Seattle: University of Washington Press.

Kitchin, E. A. 1949. *Birds of the Olympic Peninsula*. Port Angeles, Washington: Olympic Stationers.

Kozloff, Eugene N. 1976. *Plants and Animals of the Pacific Northwest*. Seattle: University of Washington Press.

———. 1983. *Seashore Life of the Northern Pacific Coast*. Seattle: University of Washington Press.

Kruckeberg, Arthur R. 1991. *The Natural History of Puget Sound Country*. Seattle: University of Washington Press.

Kuramoto, Richard, and L. Bliss. 1970. "Ecology of Subalpine Meadows in the Olympic Mountains." *Ecological Monographs*, 40: 317–347.

Larrison, Earl J., and Amy Fisher. 1976. *Mammals of the Northwest*. Seattle: Seattle Audubon Society.

———. 1981. *Birds of the Pacific Northwest*. Moscow: University of Idaho Press.

Leonard, William P., H. Brown, L. Jones, K. McAllister, and R. Storm. 1993.

Amphibians of Washington and Oregon. Seattle: Seattle Audubon Society.

Lichatowich, Jim. 1993. "The Status of Anadromous Fish Stocks in the Streams of Eastern Jefferson County, Washington," Sequim, Washington: Jamestown S'Klallam Tribe.

Lien, Carsten. 1991. *Olympic Battleground: The Power Politics of Timber Preservation*. San Francisco: Sierra Club Books.

Little, Elbert L. 1980. *The Audubon Society Field Guide to North American Trees: Western Edition*. New York: Knopf.

Lyons, C. P. and B. Merilees. 1995. *Trees, Shrubs and Flowers to Know in Washington and British Columbia*. Vancouver: Lone Pine Publishing.

Maser, Chris. 1989. *Forest Primeval: The Natural History of An Ancient Forest*. San Francisco: Sierra Club Books.

Maser, Chris, and James M. Trappe (eds.). 1984. *The Seen and Unseen World of the Fallen Tree*. General Technical Report PNW-164, Portland: USDA Forest Service and USDI Bureau of Land Management.

Maser, Chris, R. F. Tarrant, J. M. Trappe, and J. F. Franklin. 1988. *From the Forest to the Sea: A Story of Fallen Trees*. Portland: USDA Forest Service, Pacific Northwest Research Station.

Matthews, Daniel. 1988. *Cascade-Olympic Natural History: A Trailside Reference*. Portland: Raven Editions/Portland Audubon Society.

McKee, Bates. 1972. *Cascadia, the Geologic Evolution of the Pacific Northwest*. New York: McGraw Hill.

McNulty, Tim. 1990. *Washington's Wild Rivers: The Unfinished Work*. Seattle: The Mountaineers.

McNulty, Tim, and Pat O'Hara. 1984. *Olympic National Park: Where the Mountain Meets the Sea*. Del Mar: Woodlands Press.

McPhee, John.1981. *Basin and Range*. New York: Farrar, Straus, Giroux.

Moorhead, Bruce B. 1994. *The Forest Elk: Roosevelt Elk in Olympic National Park*. Seattle: Northwest Interpretive Association.

Morgan, Murray. 1955. *The Last Wilderness*. New York: The Viking Press.

Morgenroth, Chris. 1991. *Footprints in the Olympics: An Autobiography*. Fairfield, Washington: Ye Galleon Press.

Morrison, Peter. 1989. *Ancient Forests of the Olympic National Forest: Analysis from a Historical and Landscape Perspective*. Washington, D.C.: The Wilderness Society.

Nadkarni, Nalini M. 1985. "Roots that Go Out on a Limb." *Natural History*, Vol. 94.

Nash, Roderick. 1969. *Wilderness and the American Mind*. New Haven: Yale University Press.

Nehlsen, Willa, J. Williams, and J. Lichatowich. 1991. "Pacific Salmon at the Crossroads: Stocks at Risk from California, Oregon, Idaho, and Washington." *Fisheries*, Vol. 16, No. 2.

Nickerson, Roy. 1989. *Sea Otters, A Natural History Guide*. San Francisco: Chronicle Books.

Norse, Elliot A. 1990. *Ancient Forests of the Pacific Northwest*. Washington, D.C.: The Wilderness Society/Island Press.

Osborne, Richard, J. Calambokidis, and E. Dorsey. 1988. *A Guide to Marine Mammals of Greater Puget Sound*. Anacortes, Washington: Island Publishers.

Pascua, Mary Parker. 1991. "Ozette, A Makah Village in 1491." *National Geographic*, October.

Peterson, Roger Tory. 1990. *A Field Guide to Western Birds*. Boston: Houghton Mifflin Company.

Pinchot, Gifford. 1945. *Breaking New Ground*. New York: Harcourt Brace Javanovich.

Pojar, Jim, and Andy MacKinnon. 1994. *Plants of the Pacific Northwest Coast*. Vancouver, British Columbia: British Columbia Ministry of Forests/ Lone Pine Publishing.

Rau, Weldon W. 1980. *Geology of the Washington Coast Between the Hoh and Quillayute Rivers*. Olympia: Washington Department of Natural Resources Bulletin No. 72.

Ricketts, Edward, J. Calvin, and J. Hedgpeth. 1968. *Between Pacific Tides* (4th ed.). Stanford: Stanford University Press.

Robbins, Chandler S., B. Bruun, H. Zim, and A. Singer. 1983. *Birds of North America*. New York: Golden Press.

Samuels, Stephan. 1991. *Ozette Archaeological Project Research Reports*, Volume 1. Seattle: U.S. Department of Anthropology Reports of Investigations 63, National Park Service, Pacific Northwest Region.

Scheffer, Victor B. 1995. *Mammals of the Olympic Peninsula*. Olympia, Washington: Society for Northwestern Vertebrate Biology.

Sharp, Robert P. 1960. *Glaciers*. Eugene: University of Oregon Press.

Sharpe, Fred. 1992. *Olympic Wildlife Checklist*. Seattle: Northwest Interpretive Association.

———. *Olympic Peninsula Birds: the Songbirds*, unpublished manuscript, Olympic National Park.

Shelton, John S. 1966. *Geology Illustrated*. New York: W. H. Freeman and Co.

Skinner, Brian, and Stephen Porter. 1989. *The Dynamic Earth: An Introduction to Physical Geology*. John Wiley and Sons.

Smithson, Michael, and Pat O'Hara. 1993. *Olympic: Ecosystems of the Peninsula*. Helena, Montana: America and World Geographic Publishing.

Snively, Gloria. 1978. *Exploring the Seashore in British Columbia, Washington and Oregon*. Vancouver and Seattle: Gordon Soules Book Publishers Ltd./Pacific Search Press.

Steelquist, Robert. 1987. *Washington's Coast*. Helena, Montana: American Geographic Publishing.

———. 1992. *Adopt-a-Stream Foundation Field Guide to the Pacific Salmon*. Seattle: Sasquatch Books.

Stewart, Charles. 1988. *Wildflowers of the Olympics and Cascades*. Port Angeles, Washington: Nature Education Enterprises.

Stewart, Hillary. 1984. *Cedar: Tree of Life to the Northwest Coast Indians*. Vancouver: Douglas and McIntyre.

Suttles, Wayne. 1987. *Coast Salish Essays*. Seattle: University of Washington Press.

———. 1990. *Handbook of North American Indians: Volume 7, Northwest Coast*. Washington, D.C.: Smithsonian Institution.

Swan, James. 1857. *The Northwest Coast*. New York: Harper and Brothers.

———. 1974. *The Indians of Cape Flattery*. Washington, D.C.: Smithsonian Institution, facsimile reproduction: Seattle: The Shorey Book Store.

Tabor, Rowland W. 1987. *Guide to the Geology of Olympic National Park*. Seattle: Pacific Northwest Parks and Forests Association.

Tabor, R. W., and W. M. Cady. 1987. *Geological Map of the Olympic Peninsula*. U.S. Geological Survey.

Twight, Ben W. 1983. *Organizational Values and Political Power: The Forest Service Versus the Olympic National Park*. University Park: Pennsylvania State University Press.

Udvardy, Miklos. 1977. *The Audubon Society Field Guide to North American Birds: Western Region*. New York: Knopf.

Van Syckle, Edwin. 1981. *They Tried to Cut It All: Grays Harbor—Turbulent Years of Greed and Greatness*. Seattle: Pacific Search Press.

U.S. Department of the Interior, Department of Commerce, and Elwha S'Klallam Tribe. 1994. *The Elwha Report*.

U.S. Department of the Interior, National Park Service. 1995. *Elwha River Ecosystem Restoration, Final Environmental Impact Statement*.

Vermeer, Kees, K. Briggs, K. Morgan, and D. Siegel-Causey. 1993. *The Status, Ecology, and Conservation of Marine Birds of the North Pacific*. Ottawa: Canadian Wildlife Service.

Washington Department of Fisheries, Washington Department of Wildlife, and Western Washington Treaty Tribes. 1992. *Washington State Salmon and Steelhead Stock Inventory*. Olympia.

Wertheim, Anne. 1984. *The Intertidal Wilderness*. San Francisco: Sierra Club Books.

Whitlock, Cathy. 1992. "Vegetational and Climatic History of the Pacific Northwest during the Last 20,000 Years: Implications for Understanding Present-Day Biodiversity." *The Northwest Environmental Journal*, 8:5–28.

Wickersham, James. 1961. "A National Park in the Olympics. . . 1890," *The Living Wilderness*, Summer-Fall.

Wood, Robert L. 1967. *Across the Olympic Mountains: The Press Expedition, 1889–90*. Seattle: University of Washington Press.

———. 1976. *Men, Mules and Mountains: Lieutenant O'Neil's Olympic Expeditions*. Seattle: The Mountaineers.

———. 1984. *Olympic Mountains Trail Guide*. Seattle: The Mountaineers.

———. 1995. *The Land that Slept Late: The Olympic Mountains in Legend and History*. Seattle: The Mountaineers.

Wunderlich, Robert, B. Winter, and J. Meyer. 1994. "Restoration of the Elwha River Ecosystem and Anadromous Fisheries." *Bethesda* 19 (8).

Wydoski, Richard S., and R. Whitney. 1979. *Inland Fishes of Washington*. Seattle: University of Washington Press.

Yorath, C. J. 1990. *Where Terranes Collide*. Victoria: Orca Publishers.

Zwinger, Ann, and Beatrice Willard. 1972. *Land Above the Trees: A Guide to American Alpine Tundra*. New York: Harper and Row.

INDEX